A Mathematical Approach to Biology

John L. Howland
Charles A. Grobe, Jr.

Bowdoin College

D. C. HEATH AND COMPANY
Lexington, Massachusetts Toronto London

Copyright © 1972 by D. C. Heath and Company.

All rights reserved. No part of this publication may be reproduced or transmitted in any form or by any means, electronic or mechanical, including photocopy, recording, or any information storage or retrieval system, without permission in writing from the publisher.

Published simultaneously in Canada.

Printed in the United States of America.

International Standard Book Number: 0-669-75275-4

Library of Congress Catalog Card Number: 71-168977

A Mathematical Approach to Biology

PREFACE

Recent decades have witnessed a growing willingness on the part of biologists to use mathematical techniques in their discipline. Statistical analysis and modern computing machines, to name just two examples, have had an impact on virtually all regions of biology. Conversely, because the construction of mathematical models in biology has become so extremely wide-ranging in approach, many new and different areas of mathematics have become of interest to the biologist.

Hence, the field of mathematical biology has developed into something of a self-contained edifice during recent years. Several books now exist which can serve to introduce that edifice to mathematicians with biological training and vice versa. This book exists with a somewhat different, and considerably more modest, mission: to introduce undergraduate students to some examples of how elementary mathematics can be applied to biology. It is selective in its choice of topics; indeed, they were chosen mostly as examples of how a little mathematical formalism can go a long way toward improving one's understanding of a biological situation.

This book was written by a biologist and a mathematician (who, the reader will be glad to learn, are still on speaking terms). Its writing was an exercise in communication which was of benefit to both of the authors. We hope that the book will be of benefit to other people, as well.

We were aided by a number of colleagues, among whom we would particularly like to thank Dan E. Christie of Bowdoin College, for it was he who first started us thinking about the possibility of writing materials on the applications of mathematics to biology.

For their skillful preparation of the manuscript, we would like to thank Mrs. Stuart Richardson and Mrs. Merle MacDonald, the latter of whom typed and proofread the major portion of it. And finally, we are grateful to Miss Martha Allen of D. C. Heath and Company for her superb job of editing.

J. L. Howland
C. A. Grobe, Jr.
Brunswick, Maine

CONTENTS

1 Introduction

The Reluctant Mathematician 1
Biological Explanation 2
This Book 3

2 The Mathematics of Growth

The Basic Experiment 5
The Construction of the Mathematical Model 7
The Number t_e 11
The Doubling Time t_d 12
The Significance of the Growth Constant K 13
Growth in a Continuous Flow System 13
A Critique of the Model 15
More Inclusive Models for Cell Growth 16

3 Enzymes

Rates of Chemical Reactions 21
Rates of Enzyme Reactions 23
Graphs of Enzyme Reactions 27
The Use of Network Theory 31

4 Probability and Its Application

Definition of Probability   36
The Binomial Distribution   38
Deviation from the Mean Value   40
An Application to Biology   41

5 The Genetics of Populations and the Hardy-Weinberg Law

The Chromosomal Basis of Heredity   45
Mendelian Inheritance   50
Population Genetics   58
    Random Mating. The Hardy-Weinberg Law   60
    Random Union of Gametes   66

6 The Applications of the Computer to Biology

The Modern Digital Computer   69
The Program   71
The Use of Computers in Ecology   77
Computers in Physiology and Biochemistry   79

7 Some Applications of Advanced Mathematics

Calculus and Differential Equations   84
Algebra   85
Topology   86

Suggested Additional Readings   89

Index   93

A Mathematical Approach to Biology

Chapter 1

Introduction

The Reluctant Mathematician

There appears to be a view, widely acclaimed in certain circles, which holds that the skills required of the practicing biologist and the practicing mathematician are seldom found in the same individual. This view maintains, moreover, that a biologist with skill in and knowledge of mathematics will rarely find an opportunity to bring his mathematical knowledge to bear in a meaningful way upon a problem in biology. A delight in living things, a wonder at the vast diversity of life, an appreciation for the beauty to be found in nature—all of these serve to motivate scores of young people to begin a study of biology. Mathematics, on the other hand, is not everyone's cup of tea. In fact, if the truth be known, many biologists lean toward the view that mathematics is simply not for them.

In this small book, we wish to argue the antithesis of the point of view described above, for we believe that biologists, whatever preconceptions they may or may not have concerning mathematics, will have an increasing need for this subject. The primary purpose of this book, then, is to consider a number of examples of the ways in which the biologist can use mathematics to solve problems in his own field. If, at the same time, we engender a desire in the biologist to learn more mathematics, then this book shall have done more than fulfill the purposes for which it is intended.

In fairness, it must be said that there is a small, but increasing, number of biologists who use mathematics at all levels. These (we think) fortunate people call themselves (or are called by others) biomathematicians. The sum total of their work comprises the field of *biomathematics* or mathematical biology. Sometimes it appears to be mostly statistics, and sometimes it is indistinguishable from biophysics or even ecology. But the thread which appears to be common to all of these diverse subjects is a concern for and an interest in the construction and evaluation of what are called "mathematical models." In this book, we shall attempt to show by example what a mathematical model is, how one constructs and analyzes such a beast, and what one can learn from it.

Biological Explanation

Biology, as everyone knows, is the study of life—a study well known to be difficult, or we would not have been so long about it. The biologist, if he knows what is good for him, avails himself of any and every technique which stands any chance, however unpromising it may appear at first glance, of telling him something interesting about his subject. Living things—even something as small as a virus—are horribly complex objects. The biologist, therefore, should be willing to employ any technique that will provide him with a simpler way of looking at things. Mathematics, perhaps surprisingly, can be of great help in just this way.

By way of an analogy, let us turn to an example from physics. For a very long time, people have noted the effects of gravity. For instance, when one poured boiling oil from a battlement, it was observed that the oil tended to fall downwards. Indeed, until fairly recent times, mankind's knowledge of gravity could be summed up rather completely by saying that an object that is heavier than air tends to fall to the ground unless something is slid under it. To say the least, this is a purely descriptive model; it is completely inadequate to answer questions such as: If one drops an object weighing M grams from a height of H meters, how long will it take the object to strike the ground and at what velocity will it be traveling? An enormous advance in mankind's understanding of gravity came about as the result of a model constructed by Newton. He assumed that since the sun and the planets are approximately spherical in shape, their mass could be considered to be concentrated at their center. He then assumed that if, for example, you disregarded the effects of the other planets in the solar system, then the gravitational attraction between the sun and the earth is inversely proportional to the square of the distance between them. This model, together with Newton's second law of motion, gives a reasonably accurate description of gravity. Hence, through the vehicle of mathematics, it became possible to give a much more detailed description of gravity than had been possible before.

Many physicists are all too fond of pointing out that much of biology is still in a pre-Newtonian state. Although this is an extreme point of view, it is true that many biologists are just beginning to learn to ask fundamental questions in a mathematical format. And asking questions in a mathematical way amounts to constructing (or, what is often just as instructive, *trying* to construct) a mathematical system that attempts to represent some biological phenomenon.

This Book

As we have said, we shall attempt to convince you that mathematics is a useful subject for a biologist to know. We shall do so by the simple expedient of considering several examples; that is, we shall show you some instances of where a mathematical treatment of some rather simple biological information will give rise to conclusions that are useful and that are not likely to be obvious in the absence of the mathematical analysis.

For the most part, you will need to know only high school algebra. We do, however, make some use of elementary calculus. Also, we shall confine the biological discussion to topics that are of genuine interest to biologists. In striving for mathematical sophistication, it is easy to become so enthusiastic that one ends up with a situation that is either trivial or incorrect in a biological sense. For example, one of the authors had occasion to read a mathematical treatment of genetic theory which was mathematically pure and beautiful. Unfortunately, it was based on the Lamarkian† approach to genetics. The author of the article evidently overlooked the fact that Lamarkian genetic theory simply doesn't work.

There is a list of references at the end of this book. We urge you to look at it because we believe that it may actually be useful.

---

†Lamark (1744–1829) argued that characteristics of an organism that were acquired during one generation were able to be transmitted to the next, or, to put it differently, that acquired features could become incorporated into the hereditary pattern for subsequent generations. This theory has been subjected to repeated tests through the years and has been repeatedly disproven. It has even become discredited in the U.S.S.R. where, however, it lingered on into the 1950s, apparently due to ideological considerations.

Chapter 2

The Mathematics of Growth

As we noted earlier, we shall attempt to describe and to analyze certain biological phenomena which are amenable to mathematical analysis. The growth of a population of cells is perhaps one of the best known examples of such a system. The system is easy to describe biologically and requires only elementary mathematics to analyze. Furthermore, the mathematical analysis of the model will lead us to expect a certain growth pattern in the population of cells that we are studying. Because the growth pattern which we can actually observe is often in such close agreement to that which is predicted by the mathematical analysis of the model, we are naturally led to believe that the approach which we shall describe below is relevant.

The Basic Experiment

We now proceed to describe the experiment that we wish to analyze. It is important to remember that many of the details are *not* relevant to the mathematical model. After we have described the experiment in some detail, we shall, in the next paragraph, enumerate the assumptions which are important for the mathematical analysis of the biological system.

We consider a population composed of individual bacteria of the species *Escherichia coli*. These are to be grown suspended in a favorable nutrient medium. We shall grow the cells at a constant temperature—say, 37° centigrade—with constant agitation. This latter point is important because it ensures aeration of the culture (these bacteria respire). At the same time, the agitation prevents the bacteria from clumping; that is, it assures us that the culture will remain homogeneous. Furthermore, we shall assume that the flask that contains the culture is a closed system; that is, nothing more will be added to the system after the bacteria have been introduced into the growth medium.

Now, suppose that the total liquid volume of the growth flask is 10 ml. and that, initially, the contents of the flask are sterile, i.e., there are no organisms inside. We then add one ml. of medium containing a reasonable number of bacteria—say, about 10^7 cells—place the flask on a shaker, and maintain the flask at 37°C. The purpose of the experiment is to determine the number of bacteria in the flask at some later time.

But how shall we count the number of bacteria in the flask? This can be most easily done by shining a light through the medium and measuring the amount of light that is scattered by the bacteria. This, it turns out, is a very reasonable way of measuring the number of bacteria in the culture because it can be shown that the number of cells in the culture is directly proportional to the amount of light that is scattered by the medium. Finally, as a practical detail, we assume that the flask into which the bacteria have been introduced is a side-arm flask such as is shown in Figure 2–1.

A side-arm growth flask

Figure 2-1

When we wish to measure the amount of light scattered by the specimen, we simply tip the contents of the flask into the side arm and place it in the light beam of a colorimeter. This measures the amount of light that is scattered by the sample and, hence, the number of bacteria in the culture. The use of a side-arm flask has the advantage that it is not necessary to open the flask and remove a sample of the medium in order to measure its ability to scatter light. This in turn means that the flask can remain sealed for the duration of the experiment, that is, that the bacteria remain under isolated conditions. Hence, it is really quite difficult, although not impossible —we never underestimate the ingenuity of some of our students—to botch up the experiment.

The Construction of the Mathematical Model

In the paragraph above, we have described the biological experiment in some detail. Recall that the whole point of the experiment was to determine, at any given time, the number of bacteria that are in the culture. We described a method by which we could actually measure the number. *But, can we predict the number?* That is, can we write down a mathematical expression which will tell us the number of bacteria that we might expect to find in the flask at any given time? The answer is "yes" or we wouldn't have raised the question. We now proceed to show you how this can be done. The finding of the appropriate equation constitutes the "construction" of the mathematical model. In this particular case, then, our "model" is not some tangible piece of apparatus that we can manipulate and out of which will come some prediction concerning the number of bacteria in the flask. Rather, as we have said, it consists simply of a mathematical expression.

We can then compare the number of bacteria that we actually have in the flask with the number that our mathematical model predicts should be there. If the two numbers are in close agreement, then our model must have been a reasonably good one. If not, then we must modify the model, that is, we must find another mathematical expression that leads to a closer agreement between the observed and the calculated values.

In constructing the model, our first, and perhaps most difficult, task is to enumerate the assumptions that are vitally important to the experiment and to discard those that are not. For example, the kind of bacteria is not important; within limits, almost any other kind would have done just as well as *Escherichia coli*. Neither is it vital that we maintain the culture at 37°C; within reasonable limits, any temperature would suffice. What *is* important is that we maintain the culture at the *same* temperature for the duration of the experiment. Also, it is not important that the flask contains 10 ml. of nutrient; 100 ml. would be just as satisfactory. And so on.

What, then, are the critical assumptions? We shall list them below. From these assumptions, we shall proceed to write down a mathematical expression that is consistent with the assumptions. For the purposes of our model, we assume:

1. *That the system is closed; i.e., no cells are removed or added after the experiment begins.*
2. *That our measurements accurately reflect (no pun is intended) the number of cells in the population.*
3. *That external stimuli such as temperature and light will be constant for the duration of the experiment.*
4. *That new cells arise only as the offspring of cells already in the population.*

Before we attempt to analyze the consequences of these assumptions, we point out that there is nothing in any way "sacred" about them. If we had chosen a different set of assumptions, then we might have been led to a different mathematical model, which in turn might have given us a different predicted value for the number of bacteria at a given time. The closer the predicted value is to the observed value, the better, in some sense, is the model.

Let us now consider the consequences of the four assumptions. In particular, we shall expand upon the fourth one. It is certainly reasonable, for it amounts to saying that cells are factories for (among other purposes) the production of more cells. From this, it follows that the more cells there are in a population, the more factories there are and, as a consequence of this, the greater should be the rate of production. If, at any given time, a certain fixed percentage of the population is reproducing, then doubling the population will mean that twice as many cells are reproducing as before. That is, the increase in the population is proportional to the number of individuals in that population. In fact, if you think about it for a moment, you will realize that all of the assumptions that we made above are designed to insure just that. Hence, we can replace our four assumptions above by the following:

We assume that the rate at which the population is increasing is proportional to the number of members in the population.

The Construction of the Mathematical Model

We may express this more succinctly using mathematical notation. If we let N be the number of cells in the population (later we shall use the notation $N(t)$ in order to remind you that N depends upon the time t), then we may write that

(1) $$\text{Growth rate} = KN,$$

where K is a constant called the *growth constant*. This equation simply says that the growth rate is the product of N, the number of cells in the population, and K, the constant of proportionality.

Now, the growth rate is just the rate at which the population is changing at any given time; that is, it is the rate of change of population with respect to time. From the differential calculus, we have a very powerful and useful notation for this. We denote the rate of change of N with respect to t by the symbol $dN(t)/dt$. This is called the *derivative of N with respect to t*. Hence, equation (1) can be written

(2) $$\frac{dN(t)}{dt} = KN.$$

Next, we wish to describe another convention. We shall let N_0 denote the number of cells that are initially introduced into the previously sterile medium. If we agree to begin measuring time from the moment that the N_0 cells are introduced into the flask, then we can write that

$$N_0 = N(0).$$

Therefore, our mathematical model can be described very simply; it is just

(3)
$$\frac{dN(t)}{dt} = KN,$$
$$N_0 = N(0).$$

Keep in mind that N_0 and K are given. In fact, K depends upon a number of things such as the temperature, the kind of nutrient used, the type of bacteria that we employ, and so on. But if we agree to keep all of these fixed, then K is just a constant, i.e., it does not change with time. Hence, in attempting to "solve" the system (3), we seek a function $N(t)$ which has the property that it is proportional to its derivative and which also satisfies the condition, $N(0) = N_0$.

It is easy to solve (3) using elementary calculus (see any elementary calculus book). The function that we require is just

(4) $$N(t) = N_0 e^{Kt},$$

where e is the base of the so-called natural logarithms. That $N(t)$ given in (4) satisfies (3) is easy to verify, especially if you know just a bit of calculus. Since $e^0 = 1$, it follows that $N(0) = N_0$; since $de^{Kt}/dt = Ke^{Kt}$, it follows that (4) satisfies (3) as advertised. For some given time t, the number $N(t)$ predicts the number of bacteria that we should expect to find in our flask. This is shown graphically in Figure 2–2.

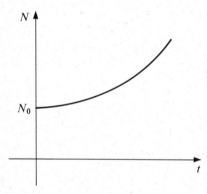

Figure 2-2

It is sometimes convenient to write equation (4) in a different form. If we let ln (N) denote the logarithm of N to the base e, then, since (4) can clearly be written as

$$\frac{N(t)}{N_0} = e^{Kt},$$

it follows that

(5) $$\ln(N(t)) = Kt + \ln(N_0).$$

This says, in plain English, that the logarithm of $N(t)$ is proportional to t. Hence, if we plot $N(t)$ against t using semilog paper, we will get a straight line, as is shown in Figure 2-3.

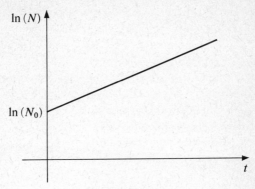

Figure 2-3

Before we proceed to deduce some of the consequences of this model, we inquire about how near our model comes to reality. In other words, if we were to run an experiment and were to plot the population against time on semilog paper, would this yield a straight line? Well, the fact is that under a wide variety of experimental conditions, we do indeed observe just this sort of behavior. Hence, we are encouraged to believe that the model we have constructed is in some sense "correct."

The Number t_e

Recall that N_0 is the number of cells that we introduced into the flask initially. We now ask: How long will it take for the number of cells to increase by a factor of e? We let t_e be the time required for the population to increase from N_0 cells to eN_0 cells. From (4), we have that

$$N(t_e) = eN_0 = N_0 e^{Kt_e}.$$

It follows that $e = e^1 = e^{Kt_e}$, and since the exponential function is strictly increasing, we may conclude that $1 = Kt_e$. Hence,

(6) $$t_e = \frac{1}{K}.$$

This, it turns out, is a useful way to determine K because t_e can be easily measured.

The number t_e was defined as the time required for the population to increase from N_0 to eN_0. Now, let N_1 be the population at some time $t_1 > 0$, i.e.,

$$N_1 = N(t_1) = N_0 e^{Kt_1},$$

and let t'_e be the time required for the population to increase from N_1 to eN_1. We wish to show that $t_e = t'_e$; that is, we wish to show that the time required for the population to increase by a factor of e is independent of the size of the population. From equation (4), we have that

$$N(t_1 + t'_e) = N_0 e^{K(t_1 + t'_e)} = eN_1 = eN_0 e^{Kt_1}.$$

It follows that

$$e \cdot e^{Kt_1} = e^{Kt_1 + 1} = e^{K(t_1 + t'_e)} = e^{Kt_1 + Kt'_e}.$$

Again, because the exponential function is strictly increasing, we have that

$$Kt_1 + 1 = Kt_1 + Kt'_e;$$

so,

$$t'_e = \frac{1}{K} = t_e.$$

This completes the argument.

The Doubling Time t_d

In a manner analogous to that given in the preceding section, we determine the time required for the population to double. Accordingly, let N_1 be the population at some time $t_1 > 0$, i.e.,

$$N_1 = N(t_1) = N_0 e^{Kt_1},$$

and let t_d be the time required for the population to increase from N_1 to $2N_1$. As before, we have that

$$N(t_1 + t_d) = N_0 d^{K(t_1 + t_d)} = 2N_1 = 2N_0 e^{Kt_1}.$$

So

$$2e^{Kt_1} = e^{\ln(2)} e^{Kt_1} = e^{Kt_1 + \ln(2)} = e^{Kt_1 + Kt_d}.$$

As before, it follows that
$$Kt_1 + \ln(2) = Kt_1 + Kt_d \Rightarrow \ln(2) = Kt_d.$$
Therefore,
$$K = \frac{\ln(2)}{t_d}. \tag{7}$$

Equation (7) provides us with still another way of obtaining the growth rate constant, K. In practice, this is the method which is most commonly used to determine K.

The Significance of the Growth Constant K

Before we go further, we wish to summarize what we have done. First of all, we described an experiment whereby we could determine the population of a certain type of bacteria as a function of time. After we had distilled away what we thought were the nonessential features of the experiment, we constructed a mathematical model that predicted the sort of growth pattern that we might expect. The model contained a constant K which we called the growth rate constant. In order to determine the number of bacteria predicted by the model to exist at some time t, we found that we must, according to equation (4), know the value of K. How, in practice, do we find it? The answer is, of course, that we can use either equation (6) or equation (7); that is, in order to determine K, we return to the actual experiment. This is a good example of the sort of constant interchange that should and does occur between the experiment, on the one hand, and the model, on the other. We should constantly regard one in the light of the other.

Remember, also, that the "constant" K represents a property of a particular population under one fixed set of conditions. If we change any of the conditions—for example, if we change the nutrient or the temperature of the nutrient—then, in general, we will change K.

Growth in a Continuous Flow System

In this section, we describe a variation of the original experiment that will lead to still another method for determining the growth rate constant K. A *chemostat* is a device that maintains a constant population of cells within a fixed volume. The cells are fed by nutrient that flows into the flask at a constant rate. In Figure 2-4 we represent a chemostat schematically; actually, it may be something of a Rube Goldberg affair.

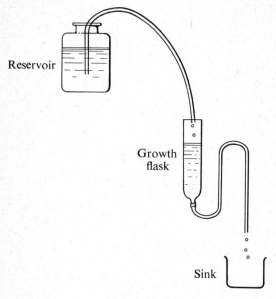

Figure 2-4

The population of cells resides in the growth flask which contains a constant volume of nutrient, V. If the rate of flow of nutrient into the growth flask is w, then, since V is constant, the flow rate of nutrient out of the flask must also be w. The flow out of the growth flask contains nutrient as well as cells.

If the rate at which nutrient and cells are flowing out of the growth flask is small in comparison to the rate at which the cells are growing, then the population of cells in the growth flask will increase. On the other hand, if the flow rate of cells out of the growth flask is greater than the rate at which the population is reproducing, then the population in the growth flask will decrease. We wish to adjust the flow rate so that the rate at which cells are exiting from the flask is exactly balanced by the rate at which the population is increasing.

Let us analyze this system. As before, the growth rate is given by

$$\frac{dN(t)}{dt} = KN.$$

The rate of dilution of cells from the growth flask is directly proportional to the flow rate w and inversely proportional to the volume V. Hence,

$$\text{Rate of dilution} = \frac{wN}{V}.$$

If the rate of dilution is the same as the rate of growth, that is, if the system is in equilibrium, then

$$\frac{wN}{V} = KN.$$

Therefore,

$$K = \frac{w}{V}.$$

Hence, we have obtained yet another way of computing K, this time in terms w and V, both of which may be easily and accurately measured.

A Critique of the Model

We have seen that our model predicts that the population of cells will increase exponentially. As we have mentioned, this prediction is in remarkably good agreement with observation in a great many cases over a short range of time. But we usually observe that when the cells are first introduced into the growth medium, they grow more slowly at first than they will a short time later. It is analogous to transplanting a tree; the tree may stop growing or actually die back until it has had a chance to regenerate some of the hair roots that it lost.

But after a time, the culture usually begins to grow in an exponential fashion as our model predicts that it should. Will this exponential growth continue indefinitely? Suppose, for example, that a bacterium reproduces once every 30 minutes. If you start with a single cell, in 30 minutes you will have 2 cells. In another 30 minutes, each of them will divide, so in 60 minutes you will have 4 cells. In 90 minutes, you will have 8 cells, in 120 minutes you will have 16 cells, and so on. Instead of counting sheep when you can't get to sleep, we suggest that you count how many bacteria you will have at the end of two days. The number is large. Now, we have nothing against bacteria—in fact, we're quite fond of some of them— but we are glad that bacteria do *not* continue to grow in an exponential fashion for an indefinite period of time.

In fact, we observe that as the environment becomes more crowded, the growth rate decreases. This should come as no great surprise, because the supply of nutrient is decreasing while at the same time the concentration of waste products is increasing. The latter, of course, may be toxic to the population of the bacteria. Hence, it is certainly reasonable to expect that the growth rate will slow down.

Also, we have assumed that each bacterium is immortal, that is, that like old soldiers, they never die. While this assumption may be reasonable during the early stages of an experiment, observations indicate that it is untenable later on.

In summary, then, what can we say about our model? Certainly, it has the advantage of simplicity, for its construction and analysis require only the most rudimentary knowledge of calculus. And, the predictions based on the model are, as we have said, in remarkably good agreement with the number actually observed in an experiment if we restrict t within limits. But, as we have also observed, the model fails to predict that the growth rate will slow down after the population begins, for example, to exhaust its nutrient supply. For this reason, we must attempt to change our model so that it takes this and other factors into consideration.

More Inclusive Models for Cell Growth

First, we wish to modify our model so that it takes account of the fact that some of the cells will die. Thus, the expression for the growth of the population becomes one for the *net growth*; that is, cells will leave the population because of death and must be subtracted from the number that are being born. Specifically, our assumption is:

> We assume that both the increase in population due to birth and the decrease in population due to death are proportional to the population.

In this case, then, we may write that

$$\frac{dN(t)}{dt} = KN - PN = (K - P)N,$$

where N is the number of cells in the population, K is the growth rate constant, and P can be called the *death rate constant*. But, if K and P are constants, then we can replace $K - P$ by a new constant, say C, and we are right back where we started. That is to say, this model, mathematically speaking, is not one bit different from the previous one.

In an actual experiment, we observe, in fact, that K and P are not constants. They change with time. We elaborate upon this point.

In this model, we shall assume that the population tends toward some maximum value determined by space and nutrient. As the population increases, its *rate of growth* will get smaller as the number of cells gets closer and closer to this maximum value. Specifically:

> We assume that the rate of growth of the population is proportional to the difference between the maximum population which the medium can support and the population at any given time.

We may write, therefore, that

(8) $$\frac{dN(t)}{dt} = K(N_{max} - N),$$

$$N_0 = N(0),$$

where K is constant and N_{max} is the maximum population which can be supported. Equation (8) may again be solved using techniques from elementary calculus. We obtain

(9) $$N(t) = N_{max} - (N_{max} - N_0)e^{-Kt}.$$

Notice that
$$\lim_{t \to \infty} e^{-Kt} = 0;$$
hence,
$$\lim_{t \to \infty} N(t) = N_{\max}.$$

As usual, we must expose our model to the cold realities of life. Just how well, then, does the population that is predicted by equation (9) agree with that observed by experiment? In order to answer this question, we observe that equation (9) predicts, because of the factor e^{-Kt}, that the rate of growth will always decrease. But, we know that over much of the growth period, the population is increasing. Hence, while this model predicts $N(t)$ with fair accuracy for large values of t, it is of little worth for intermediate values of time. Hence, one would like a model that incorporates the desirable features of both of the previous models. We simply proceed as follows:

> We assume that the rate of growth of the population is proportional both to the size of the population and to the difference between the maximum population which the medium can support and the population at any given time.

We may then write that

(10) $$\frac{dN(t)}{dt} = KN(N_{\max} - N),$$

$$N_0 = N(0).$$

The solution to (10) is

(11) $$N(t) = \frac{N_{\max} N_0}{N_0 + (N_{\max} - N_0)e^{-N_{\max}Kt}}.$$

Note, as before, that
$$\lim_{t \to \infty} N(t) = N_{\max}.$$

The model that we have just constructed is the last that we shall construct for population growth. Although the model has been widely used as an approximation to population behavior, its applicability is limited. As such models go, it is primitive.

Other, more elegant, more complicated, more useful models have been formulated. Rather than provide you with a complete catalogue of such models, we mention some of the questions which these models will hope to answer. For example, some are concerned with internal properties of growing cells. They attempt to deal with the intercellular processes, as well as the extracellular environment, that limit the growth of a population. This is different from the approach we adopted above. We viewed the cell simply as some sort of growing machine whose inner workings are irrelevant to the discussion.

Additional factors may be considered. In most populations of cells, cell division is not in synchrony; i.e., the cells are dividing somewhat at random. There are a number of difficult and interesting questions that may be asked about the degree of synchrony that can and does exist in a population and how it may be measured.

Furthermore, in a population where death is an event which occurs frequently enough to be significant, cells have an average lifetime. We may then inquire into the age distribution of the population. In populations where reproduction occurs only during a portion of the individual's lifespan, the age distribution becomes of great significance in determining the growth rate itself. While this sort of consideration is obviously important in determining the growth rate of higher organisms (including, we presume, the reader himself), it may also be important when one is studying populations of single-celled beings. This comes about because cells that are mortally ill are not as likely to grow normally, and hence, to reproduce, as are those in their prime.

In summary, it should be clear that most of the really interesting and significant questions that naturally arise when one wishes to study population growth will not yield much ground if one's approach is based entirely on intuition or common sense. On the other hand, the same questions are sometimes easily answered if one has recourse to some quite simple mathematical techniques.

Chapter 3

Enzymes

Enzymes are catalytic proteins that increase the rate of essentially all chemical reactions in living organisms. Their study is central to much of physiology and all of biochemistry. It also provides an excellent example of the important conclusions that may be drawn from simple mathematical arguments. The properties of enzymes which are most significant and which concern us here include their high degree of specificity, their enormous catalytic power, and their large molecular weight. By *specificity* we mean the ability of an enzyme to "recognize" its proper reaction partner. For instance, an enzyme that catalyzes the breakdown of one sugar is unlikely to work with others. By *catalytic power* we refer to the enormous increase in the rate of a reaction which occurs when a small amount of the appropriate enzyme is added. Moreover, the enzyme, being a true catalyst, is not used up during the course of the reaction. Hence one molecule of enzyme can preside over the alteration of many molecules of reactant. Finally, the enzyme is very large when compared with the reactant molecules. Hence, the reactant molecules may be thought of as interacting with a portion of the enzyme surface at the so-called *active site*.

Rates of Chemical Reactions

In this chapter, we will consider the rates of enzyme-catalyzed reactions and we will see how simple assumptions about enzyme action lead, via modest algebra, to quite specific and elegant conclusions. At the same time, the usefulness of certain kinds of graphs will become apparent. First, however, it is necessary to establish some conventions which apply to the rates of reactions in general (no matter whether enzyme-catalyzed or not). In the first place, suppose we consider a hypothetical reaction involving a single sort of molecule, A, which is transformed into a second one, B. We draw the reversible reaction

$$A \rightleftharpoons B.$$

We will assume that the rate of the reaction in the direction of B, which we denote by dA/dt, is proportional to the concentration of A. So, if we denote the concentration of A by $[A]$, then we can define a proportionality factor (called the *rate constant*) k via the equation

(1) $$\frac{dA}{dt} = k[A].$$

Equation (1) is valid if we assume that the reaction goes only in the direction of B.

Under conditions in which equation (1) is valid—for example, when the concentration of A is within reasonable limits—then a doubling of $[A]$ will imply a doubling of the rate of the reaction.

We do not necessarily mean to imply in (1) that only one molecule is involved in the reaction. When we wish to call your attention to the fact that more than one kind of molecule is involved, we will write something like

$$A + B \rightleftharpoons C + D.$$

Equation (1) then takes the form

(2) $$\frac{d([A] + [B])}{dt} = k[A][B].$$

In the special case where two molecules of the same thing react together, we write

$$A + A \rightleftharpoons B.$$

In this case, equation (2) becomes

$$\frac{d[A]}{dt} = k[A]^2.$$

Now, most reactions do not go in one direction only, but instead are reversible. One therefore has rate constants for both the forward and the reverse reactions. We write

$$A \underset{k_2}{\overset{k_1}{\rightleftharpoons}} B,$$

where k_1 is the rate constant from A to B and k_2 is the rate constant from B to A.

From the discussion above, the reader might be led to believe that a reaction will always proceed either from A to B or from B to A. That is, if we assume the former, then the reaction will stop only when all of A has been used up. Such is not the case. Instead, the reaction reaches what is called *equilibrium*. The idea is simply that if we allow the reaction to proceed long enough, then the ratio of A to B will eventually become constant, independent of the initial concentrations of A and B. We can also think of equilibrium as being the state when the rate of the forward reaction equals that of the reverse reaction. Hence, one can define the so-called equilibrium constant K_{eq} by the equation

(3) $$K_{eq} = \frac{[B]_{eq}}{[A]_{eq}},$$

where $[A]_{eq}$ and $[B]_{eq}$ are the concentrations of A and B, respectively, when equilibrium is attained.

The equilibrium constant K_{eq} may also be given by the equation

(4) $$K_{eq} = \frac{k_1}{k_2}.$$

The reader should attempt to convince himself that equations (3) and (4) yield equivalent definitions of K_{eq}.

Rates of Enzyme Reactions

With these definitions in mind, we return to a consideration of enzymic reactions. First of all, we must be more explicit about what we mean when we say that enzymes are catalytic. A catalyst, remember, is a substance that increases the rate of a reaction without itself being transformed by the reaction. This perhaps sounds like magic, but what actually occurs is the following: The catalyst reacts with a substance A to form a new substance B which in turn breaks down into C and the catalyst. Hence, we can write

$$A + \text{catalyst} \rightleftharpoons B \rightleftharpoons C + \text{catalyst}.$$

Thus, catalysts in general and enzymes in particular play a cyclic role and are continuously regenerated so that they can react again. Accordingly, the reason that a catalyst speeds up the reaction is that it provides an intermediate complex that is more reactive, and hence more rapidly able to form products, than the original reactants alone. Let us now apply this cyclic pattern to an enzyme-catalyzed reaction. If we let E denote the enzyme, let S denote the substance, called the *substrate*, with which E specifically reacts, and let ES denote the substance which is formed by the reaction of E with S, then we can write

$$E + S \underset{k_2}{\overset{k_1}{\rightleftharpoons}} ES.$$

Now, ES breaks down—this is the whole idea behind an enzymic reaction—to form a product, call it P, plus free enzyme. Hence, the entire reaction can be written

$$E + S \underset{k_2}{\overset{k_1}{\rightleftharpoons}} ES \rightarrow P + E.$$

So, ES can break down to give either P plus enzyme or S plus enzyme.

In order to simplify the discussion which follows, we will assume that the final reaction is one-way, that is, that it proceeds according to the scheme

$$ES \rightarrow P + E.$$

This assumption is consistent in a great number of cases with what is actually observed in the laboratory.

Now, we will introduce a second assumption which, like the first, will simplify the calculations that follow and is, at the same time, consistent with laboratory observations. It is called the *steady-state assumption*. It not only forms a basis for understanding enzyme action, but also provides us with insight into a number of other physiological and ecological processes as well. In the case of the reaction sequence above, the assumption is: *When the reaction is well* under way, *the concentration of ES must be constant*. This is a very reasonable assumption; it simply requires that the rate at which *ES* is formed must be equal to the rate at which it breaks down. In addition it is clear that the rate of the whole reaction must equal the rate of any segment of it—for example, the portion between *ES* and *P*. For otherwise, *ES* would build up and violate the steady-state assumption.

Next, we introduce notation. Specifically, we let

e = the total enzyme concentration,
p = the concentration of *ES*, the enzyme-substrate complex,
s = the concentration of the substrate.

According to laboratory observation, the substrate concentration will be very large when compared with that of the enzyme. Hence, the amount of *S* in the form of the complex can be ignored. On the other hand, one cannot ignore the amount of enzyme that forms a part of the complex at any given time. Hence, the concentration of free enzyme is actually $e - p$. The steady-state assumption requires that the

Rate of formation of *ES* = Rate of breakdown of *ES*.

Using the notation above and the definitions of rate constants implicit in equations (1) and (2), we can write

(5) $$k_1(e - p)s = k_2 p + k_3 p.$$

This gives

$$k_1(es - ps) = (k_2 + k_3)p,$$

or

$$es - ps = \left(\frac{k_2 + k_3}{k_1}\right) p.$$

Solving for *es* gives

$$es = ps + \left(\frac{k_2 + k_3}{k_1}\right) p,$$

or

$$es = p\left(s + \frac{k_2 + k_3}{k_1}\right).$$

Solving for p yields

(6) $$p = \frac{es}{\dfrac{k_2 + k_3}{k_1} + s}.$$

Hence, equation (6) gives an expression for the concentration of the enzyme-substrate complex in terms of e, s, and the three rate constants. The quantities e and s are both known by the experimenter. We shall deal shortly with the three rate constants.

We now return to the steady-state assumption. Recall that we require that each segment of the reaction must go at the same rate as the overall reaction. If this is the case, then the reaction that yields P will proceed at the same rate as the entire reaction. Since k_3 is the rate constant associated with this reaction, we have from equation (1) that

(7) $$v = \frac{dP}{dt} = k_3 p,$$

where we use v to represent dP/dt, the rate of the reaction. If we substitute the expression for p given by equation (6) into (7), we obtain

(8) $$v = \frac{k_3 es}{\dfrac{k_2 + k_3}{k_1} + s}.$$

Equation (8) can be simplified as follows. We define v_{max} to be the limit of v as the substrate concentration s becomes large. Under these conditions, virtually all of the enzyme is in the form of the enzyme-substrate complex. (Can you see why?) So, as s gets very large, p approaches e and v approaches v_{max}. Hence, when the concentration of substrate is high, we obtain from equation (7) that the difference between v_{max} and $k_3 e$ is small: Hence, we write

$$v_{max} = k_3 e.$$

If we substitute this expression into (8), we obtain the very useful result that

(9) $$v = \frac{v_{max} s}{\dfrac{k_2 + k_3}{k_1} + s}.$$

Finally, because the three rate constants are difficult to determine experimentally, we lump them together and define the so-called Michaelis† constant K_m by the equation

$$K_m = \frac{k_2 + k_3}{k_1}.$$

Equation (9) then takes the form

(10) $$v = \frac{v_{max}s}{K_m + s}.$$

Hence, we have obtained an expression for the rate of an enzymic reaction. We must now devise a method for determining the constants for a given enzyme.

Graphs of Enzyme Reactions

Equation (10) is clearly not linear in v and s; that is, it is not of the form $Av + Bs + C = 0$ for constants A, B, and C. In fact, if we plot v against s, we obtain something which looks like the curve in Figure 3-1. This is, in fact, just a hyperbola. As s becomes large,

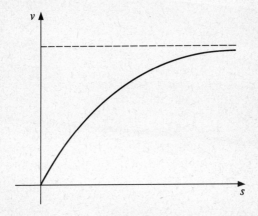

Figure 3-1

---

†For the origin of this constant and its interpretation, see any textbook of biochemistry.

we see that v approaches v_{max}. Hence, one could hope to obtain v_{max} from just such a graph. Unfortunately, it is often experimentally difficult to obtain reliable values of rate at high substrate concentrations. If, however, one can obtain v_{max}, then it is an easy matter to find K_m. In order to see why this is the case, we observe that K_m turns out to have the units of concentration and, in fact, to represent the concentration of substrate required to give a rate of $v_{max}/2$. Why is this so? If we substitute $v_{max}/2$ into equation (10), we obtain

$$\frac{v_{max}}{2} = \frac{v_{max} s}{K_m + s}.$$

Since $v_{max} \neq 0$, we have that

$$\frac{1}{2} = \frac{s}{K_m + s}.$$

This gives, then, that

$$K_m = s.$$

This is of limited practical value because, as we have said, it is usually difficult to obtain accurate values for v_{max} from a graph such as Figure 3–1.

Now, although (10), as we have seen, is not linear in v and s, it is linear in $1/v$ and $1/s$. In fact, solving (10) for $1/v$ gives

(11) $$\frac{1}{v} = \frac{K_m}{v_{max}} \cdot \frac{1}{s} + \frac{1}{v_{max}}.$$

So, if we graph $1/v$ against $1/s$, we will obtain a straight line as shown in Figure 3–2. Let us determine the intercepts, that is, the values where the line crosses the two axes. First, if we let $1/v$ approach zero, we obtain from (11) that

$$\frac{K_m}{v_{max}} \cdot \frac{1}{s} + \frac{1}{v_{max}} = 0.$$

Solving for $1/s$ gives

$$\frac{1}{s} = \frac{-1}{K_m}.$$

Graphs of Enzyme Reactions

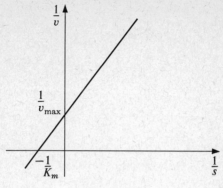

Figure 3-2

In order to obtain the other intercept, we let $1/s$ approach zero. In this case, we obtain

$$\frac{1}{v} = \frac{1}{v_{\max}}.$$

Figure 3-2 shows these values.

It is therefore possible, using a graph such as is shown in Figure 3-2, to obtain values for K_m and for $v_{\max}$ with considerable precision because it is not necessary to use large values of s.

A number of other linear graphs have been employed in order to determine the constants K_m and $v_{\max}$. Two of these are shown in Figure 3-3. The reader is invited to show that the intercepts are as given, using an analysis similar to the one which we gave for Figure 3-2.

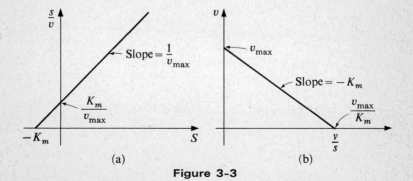

Figure 3-3

The Use of Network Theory

We have seen the importance of different graphs in obtaining quantitative information about enzyme reactions. Indeed, the same methods that we described above are of interest because they also enable us to reach more general conclusions about such matters as enzyme inhibition.

In this section, we briefly describe how another kind of graph can help us to understand something about more complicated enzyme systems.

For instance, consider the case where an enzyme reacts, not only with its substrate, but also with a modifier molecule. The latter might well be, for example, some kind of inhibitor of the net reaction. Then, if E stands for enzyme, S for substrate, M for modifier, and P for product, we may write the following:

1. $$E + S \underset{K_1}{\overset{K_2}{\rightleftharpoons}} ES \overset{K_9}{\to} E + P$$

2. $$E + M \underset{K_8}{\overset{K_7}{\rightleftharpoons}} EM$$

3. $$ES + M \underset{K_3}{\overset{K_4}{\rightleftharpoons}} EMS \overset{K_{10}}{\to} EM + P$$

4. $$EM + S \underset{K_6}{\overset{K_5}{\rightleftharpoons}} EMS \overset{K_{10}}{\to} EM + P.$$

Note that, as in the case of the previous discussion, we assume that the reactions leading to P are irreversible. Also, the complexity of the situation comes from the ability of ES to react with M and that of EM to react with S, both producing the triple complex EMS. It is clear that such situations are rather difficult to keep straight in one's mind (and this is not very complicated, as such things go). A very useful way to sort things out is by means of a network (also called a linear graph).

Let us now draw a simple graph showing the possible relationships. The graph will actually represent the concentrations of the different substances. Let F_1 be the concentration of the enzyme, E; let F_2 be that of ES; let F_3 be that of EM and let F_4 be that of EMS. The graph showing the interrelation of these is given in

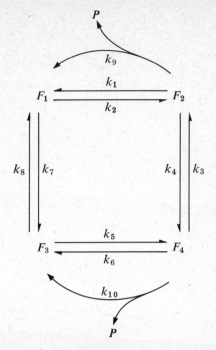

Figure 3-4

Figure 3–4. It should be clear to the reader that there is a reason for representing the reactions in this fashion. By appealing to the graph, we hope that the relationships will become clearer and the basic symmetries will become more apparent. In addition, the basic rules of network theory allow us to simplify things further and, often, to obtain an exact algebraic representation of the set of events. Some of the rules and manipulations that are possible follow.

We will call the "corners," which represent the different states of the enzyme, the *nodes*. The nodes are connected by *branches* which indicate transition between the states. Finally, we define a *path* as any sequence of branches, going in one direction, where no node is encountered more than once. An example of a branch would be from F_1 to F_2, while a path would be from F_1 to F_2 to F_4 to F_3. Obviously, a path cannot form a loop.

In order to use a linear graph to obtain enzyme rate expressions, it is first necessary to assign values for each branch. These are, in fact, obtained in terms of the different rate constants (Figure 3-4). They are called the *branch values* of the graph. These, in turn, may be multiplied together to yield *path values*. For example, in Figure 3-4 a possible path value between F_2 and F_3 would be $k_4 k_6$, and an alternate one would be $k_1 k_7$. Then the path values may be further combined (by multiplication) until an expression is obtained for all of the branches of the graph which touch each of the nodes. From this, one may then obtain expressions for the velocity of the net enzyme reaction. The technique for doing so is based on the theory of electrical networks and is beyond the scope of this book. An interested reader may find out more by reading the original paper by Volkenstein and Goldstein† that describes this approach to enzyme problems.

It should be added that a special advantage of this point of view is the number of ways that a graph may be simplified by simple operations which take note of basic symmetries in the events under consideration. For example, parallel branches may be merged, with their values being added. Finally, especially complex reaction mechanisms may be expressed as graphs in order to design an actual electrical analog which, in turn, may shed light on the problem. This amounts to substituting an electrical network for one composed of a sequence of chemical reactions, a tactic of wide application in the case of analog computers.

---

†Volkenstein and Goldstein, *Biochimica Biophysica Acta*, **115**, pp. 471–477 (1966).

Chapter 4

Probability and Its Application

Imagine that we have set up an experiment in order to measure some quantity associated with a biological system. For example, we might be trying to measure the oxygen intake of a certain strain of bacteria as a function of the temperature. Because biological systems are usually quite complicated, we would undoubtedly find that our measurements produced a spread of values, where the extremes occur rarely and where values somewhere near the middle occur most frequently. These variations in the measurement are sometimes attributed to a sort of willful spontaneity on the part of living organisms. It is often asserted that such variability does not occur in the so-called "hard sciences" such as physics and chemistry, but rather, is a characteristic of biology. If you know something about 20th century physics, you will realize that this is not the case. Thermal physics, for example, is based upon the statistical properties of collections of molecules, and measurements of such physical systems will produce a distribution of values similar to, but often narrower than, a biological system.

Whenever variations in one's measurements occur, they imply an element of randomness in the phenomena that one is attempting to measure. Or, to be more precise, the experimenter *hopes* that any variations that he observes in his measurements occur as the result of a randomness in the system that he is attempting to measure rather than as the result of the measurement itself. He must often work hard to insure that this latter possibility is minimized. Also, the experimenter would like to be able to predict the expected value of some future measurement. This is where he must have recourse to a subject called probability and statistics.

Let us consider a simple example. Suppose that we had measured the nose length of a number of people. Suppose that the shortest nose we found was two centimeters and the longest one meter. Now, if someone were to ask us for a single statement about the length of the human nose, we could reply in a number of ways. We might, for example, simply elect to take the average of the minimum and maximum values and say, therefore, that the typical human nose is just over half a meter in length. Or, we might take the average of all of the values that we observed; that is, we might add together all of the lengths and divide by the number of observations. Or, we might arrange our data in order of increasing size and choose the one in the middle, that is, the one such that there are as many noses bigger than it as there are smaller than it. All of these are reasonable things to do, but some are more reasonable than others.

As is often the case, a picture will help. Along the horizontal axis, we plot the lengths that we observed; along the vertical axis, we plot, for a fixed value of the length, the number of noses that we observed of that length. Then we fill in the points with a smooth curve. We would, if we measured enough noses, obtain a curve that looks something like the one in Figure 4-1. Looking at the graph, we then might say that the typical nose is about four centimeters long because we observed more noses of about that length than of any other. We could also, by looking at the graph, obtain information that was not available to us before, namely, that noses one meter long occur quite infrequently. (We hope that a quick look in the mirror will convince you of this.)

In setting up such a graph, it is important that we record *all* of

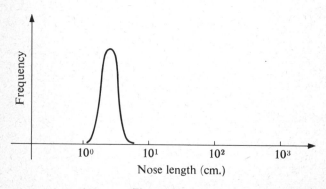

Figure 4-1

the measurements that we obtain. It is sometimes tempting to ignore some of the more scattered results on the grounds that they "must" be incorrect. To do so is not to play according to Hoyle.

There are two numbers that we will associate with a distribution such as that shown in Figure 4-1. If, as we described above, we enumerate all of our measurements in either ascending or descending order of magnitude, then we define the *median value* as the value in the middle, that is, the value that has as many numbers above it as below it. For example, suppose that we measured the lengths of just five noses and obtained the following data: 2, 3.8, 4, 7.2, and 7.5. Then the median value is 4.

The *mean value* is the average of the values; that is, it is the sum of all of the values divided by the number of values. If, again, we record the values 2, 3.8, 4, 7.2, and 7.5, then the mean value is 4.9. We can symbolize this procedure as follows: If we obtained n measurements, say $x_1, x_2, x_3, \ldots, x_n$, then the mean value, often denoted by $\bar{x}$, is given by the equation

(1) $$\bar{x} = \frac{1}{n}(x_1 + x_2 + x_3 + \cdots + x_n).$$

As can be seen from the above example, the mean and median are usually not the same.

After we have collected our data and have drawn a graph such as that shown in Figure 4-1, then we may begin to ask questions such as: "How likely is it that a subsequent measurement will yield a particular value?" and "How large a sample does one need in order to be able to draw meaningful conclusions from the data?" In an attempt to get at least a partial answer to these questions, we embark upon a brief introduction to a subject called probability and statistics. This subject represents a way of analyzing a collection of data or, to say it differently, statistics and probability provide us with techniques by which we may make reasonable generalizations from a collection of data. Because a model will often predict a single value for some quantity that we are trying to measure and an experiment will yield a number of different observed values, probability and statistics may provide us with a means of evaluating the model. If the number predicted by the model is close to the mean, say, of the observed values, then our model may be a good one.

Definition of Probability

Let us summarize what we have learned so far. If we are conducting an experiment on some biological system in order to measure some quantity associated with that system, we know that we may expect some randomness in the measurements that we obtain. Hence, we are never sure *in advance* what value we will obtain next. But neither are we completely in the dark because, if we have carried out enough measurements previously, we have some idea of what value we are likely to obtain. And as we collect more and more data, we become better informed about the results that we can reasonably expect to obtain from future measurements.

We shall define the probability of an event in terms that are more useful than elegant, but in terms that should be meaningful to anyone. When we refer to the *probability of an event*, we mean the frequency with which that event occurs in a very large number of trials. This definition binds the probability of an event to measurement or to the possibility of measurement.

Now, an event can't be more probable than certain nor less probable than impossible. So, if the probability of an event is given by P, then we establish the convention that

$$0 \leq P \leq 1;$$

that is, if an event is impossible, we assign it the probability 0 and if an event is certain, we assign it the probability 1. So, to paraphrase Ben Franklin, we can say that the probability of your escaping either death or taxes is, with the present state of technology, zero.

We define the *frequency F of an event x in n measurements* by the equation

$$F = \frac{x}{n}.$$

For example, if we toss a coin 100 times and get heads 43 times (an experiment which one of the authors actually conducted in his high school history class), then we say that the frequency of heads is $43/100 = 0.43$. We note that F approaches P as the number of measurements increases. So, in other words, when the number of measurements increases, the frequency becomes an increasingly reliable measure of the probability; that is, the difference between the frequency and the probability decreases. Since the number of measurements that one can make is obviously finite, the numbers F and P are not, in general, identical. One then has to make a decision concerning the amount of deviation he is willing to accept between the two numbers. This, in turn, requires a more formal approach.

That is to say, we must know something about the events themselves. For instance, one might first ask if two or more events are mutually exclusive; that is, does the occurrence of one event exclude the occurrence of the other(s)? For example, if we toss a coin, then either we get heads or we get tails, but we can't get both. Hence, the two events are mutually exclusive.

If E_1 and E_2 are two events, then by the notation $E_1 \cup E_2$ we mean that either E_1 will occur or that E_2 will occur, or both. If we denote the probability that E_1 will occur by $P(E_1)$, the probability that E_2 will occur by $P(E_2)$, and the probability that E_1 or E_2 or both will occur by $P(E_1 \cup E_2)$, then the equation

(2) $$P(E_1 \cup E_2) = P(E_1) + P(E_2)$$

provides us with a way of determining when two events are mutually exclusive; that is, E_1 and E_2 are mutually exclusive if and only if they satisfy equation (2). This definition can be extended to n events in an obvious way. So, for example, if we toss a coin and let E_1 denote the appearance of heads and E_2 denote the appearance of tails, then certainly $P(E_1 \cup E_2) = 1$ because the probability that we will get either heads or tails is certainly 1. We are, of course, interested in the probability that heads but not tails will occur. Over the years, a considerable amount of empirical evidence suggests that the probability of our getting heads if we toss a coin is 1/2.

If we toss a coin or roll a die, we assume that the coin or die is "honest," that is, that it is not loaded, and that the roll or toss is random. Hence, the successful application of statistics and probability to a model requires a considerable understanding of the system being studied. Despite the wishes of some, the model is not a substitute for such understanding.

It is also necessary to consider the probability of events that are not mutually exclusive. For example, we might wish to roll a die and inquire into the probability that we will get either an even number or a number that is divisible by 3. So, if we let E_1 denote the occurrence of an even number and E_2 denote the occurrence of a number which is divisible by 3, then we see that E_1 and E_2 are not mutually exclusive because the number 6 satisfies both E_1 and E_2.

Now, if E_1 and E_2 are any two events, then by the notation $E_1 E_2$ we mean that *both* events occur. In this case, we write

(3) $$P(E_1 E_2) = P(E_1) P(E_2),$$

where, as before, $P(E_1)$ represents the probability that event E_1 will occur, $P(E_2)$ represents the probability that E_2 will occur, and $P(E_1 E_2)$ represents the probability that $E_1 E_2$ will occur. Indeed, equation (3) may be used to define what one means by two events being independent. That is, two events E_1 and E_2 are said to be independent if and only if they satisfy equation (3). This definition may be extended in an obvious way to n events.

The Binomial Distribution

Suppose that we have a number of independent events to consider and we wish to know the probability that each will occur. There is a rule for doing this which, upon some reflection, seems very reasonable. The general rule is simply this:

> *The probability of a given result is equal to the number of ways that the result can occur divided by the total number of possible results.*

For example, suppose that we throw two dice and wish to calculate the probability of getting two numbers whose sum is three. In the following table, we list across the top the numbers appearing on the first die, and down on the left-hand side, the numbers appearing on the second die. On the intersection of the row and column of each summand, we list the sum. In this way we obtain Table 1. Since 3 can be gotten in two ways and there are 36 possibilities in all, we see that the probability of obtaining a 3 with a roll of two dice is 2/36. On the other hand, the probability of obtaining a 7 is 6/36.

Now suppose that we conduct n independent experiments, where we know that we may expect result A with probability p. We now wish to compute the probability P_k that we will obtain event A

Table 1

	1	2	3	4	5	6
1	2	3	4	5	6	7
2	3	4	5	6	7	8
3	4	5	6	7	8	9
4	5	6	7	8	9	10
5	6	7	8	9	10	11
6	7	8	9	10	11	12

in k of the n measurements, where $k < n$. For example, we might wish to throw two dice 10 times and ask what the probability is that a 6 will appear exactly 3 times in the 10 rolls of the dice.

There is a formula which we borrow from probability which allows us to compute this. The probability P_k may be found by the formula

(4) $$P_k = \binom{n}{k} p^k (1-p)^{n-k},$$

where $\binom{n}{k}$ is the so-called binomial coefficient given by

$$\binom{n}{k} = \frac{n!}{k!\,(n-k)!}.$$

Let's return to our example. We know that the probability that a 6 will occur in one roll of the dice is 5/36. So $p = 5/36$. Hence $1 - p = 31/36$. If we roll the dice 10 times, then $n = 10$, and if we want the probability that 6 will occur exactly 3 times in the 10 rolls, then $k = 3$. So, according to equation (4), the probability that we will throw a 6 exactly 3 times in 10 rolls of the two dice is

$$P_3 = \frac{10!}{3!\,7!} \left(\frac{5}{36}\right)^3 \left(\frac{31}{36}\right)^7.$$

What is the probability that you would throw a 6 exactly 7 times in

10 rolls of two dice? (You now have enough knowledge to become your neighborhood's top crapshooter.)

Deviation from the Mean Value

Equation (4) allows us to compute the probability of a "compound" event if we know the probability of each of its components. But biological systems are usually too complicated to enable us to obtain the probability of the components. Hence, we must bear off on a different tack.

Now, we will frequently find ourselves in the position of having obtained a number of measurements on some biological system. These measurements will not usually be identical but will be of similar magnitude. We will then be interested in obtaining the mean of these measurements, which we can do by using equation (1), and some measure of the "scatter" of the measurements about the mean. For a set of measurements can have the same mean value but have quite different amounts of scatter. Since the amount of scatter is important in determining the number of measurements that represents an adequate sample, it is important that we devise a way of measuring the amount of scatter.

So, suppose that we have taken n measurements of some system and have obtained the values $x_1, x_2, x_3, \ldots, x_n$. From equation (1), we know that the mean value of these n measurements is given by

$$(1) \quad \bar{x} = \frac{1}{n}(x_1 + x_2 + \cdots + x_n);$$

i.e., $\bar{x}$ is just the average of the numbers $x_1, x_2, \ldots, x_n$. We can easily compute the difference between each of the measurements and the mean, and, since we are interested in the distance between each measurement and the mean rather than the difference itself, we consider the absolute value of the difference, i.e., $|x_i - \bar{x}|$. Now, if we take the average of these distances, that is, if we let

$$(5) \quad M = \frac{|x_1 - \bar{x}| + |x_2 - \bar{x}| + \cdots + |x_n - \bar{x}|}{n},$$

then M gives us some measure of the scatter in the n measurements $x_1, x_2, \ldots, x_n$. We call M the *mean deviation*.

Now, the mean deviation is a perfectly useful measure of the scatter but is sometimes laborious to compute. There is an alternate method for measuring the scatter in a set of measurements that we outline as follows. Suppose that the probability of obtaining measurement x_i is given by p_i. We define the *variance* of the n measurements by the equation

$$V = (x_1 - \bar{x})^2 p_1 + (x_2 - \bar{x})^2 p_2 + \cdots + (x_n - \bar{x})^2 p_n.$$

We then define the *standard deviation*, Q, as the square root of V; that is, the standard deviation is given by

$$Q = \sqrt{(x_1 - \bar{x})^2 p_1 + (x_2 - \bar{x})^2 p_2 + \cdots + (x_n - \bar{x})^2 p_n}.$$

As we said, Q is often easier to work with in practice than is M.

An Application to Biology

We now consider an application that is of considerable interest to biologists. Suppose that we wish to determine the rate of uptake into living tissue of some radioactive material. We suppose that we can determine the amount of radioactive material that is taken in by the tissue by monitoring the radiation produced by the radioactive decay of the substance. If the material taken in by the tissue has a very long half-life when compared with the duration of the experiment, then the rate of decay (disintegrations per minute) is directly proportional to the amount of material present. On the other hand, if the half-life is short in comparison to the duration of the experiment, it is a simple matter to compensate for the loss due to decay.† One must assume, of course, that the efficiency of the measuring apparatus is constant. Hence, a given number of disintegrations will always give rise to a constant number of counts.

---

†Decay is governed by an equation quite similar to the one for growth that we considered in Chapter 2. Since the rate of decay will be proportional to the number of atoms present, and since this number decreases as decay goes on, we find that the equation which describes decay is

$$N(t) = N_0 e^{-kt},$$

where $N(t)$ is the number of atoms remaining at time t, N_0 is the number of atoms at time $t = 0$, and k is a proportionality constant called the "decay constant."

Now, the disintegration of a radioactive substance is a random phenomenon. We can predict the number of counts per minute (CPM) with considerable accuracy if the experiment lasts for a sufficiently long time, but we cannot say exactly when the next count will occur. In other words, the disintegrations are irregularly spaced but the longer that we observe them, the less significant the irregularity becomes.

The need for a large number of measurements can be seen by considering some data that were collected in an actual experiment to measure the amount of tritium (^3H) that had been previously incorporated into cellular DNA. First, the material was counted for 0.1 minutes for 4 separate intervals. The individual counts are given in Table 2 together with the deviations from the mean. In this

Table 2

x	170	188	163	196
$x - \bar{x}$	9	9	16	17

case, the mean value of the four counts is 179 and the mean deviation is 12.75. Because such results are customarily given in units of counts per minute, one obtains a mean count of 1790 CPM and a mean deviation of 127.5 CPM. This represents an error of over 7%—an error that might be within acceptable limits for some purposes but not for others.

In order to illustrate the influence that a larger sample can have on the mean deviation, we counted the same sample again. But instead of counting it just for 0.1 minutes, we counted it for 2 minutes, i.e., for 20 times as long. We obtained the results given in Table 3.

Table 3

x	4023	3995	4034	3993
$x - \bar{x}$	12	16	23	18

In this case, the mean value of the four measurements is 4011 and the mean deviation is 17.25. Note that although the number of counts is about 20 times as large as those given in Table 2, the mean values in each case are about the same. When converted to a time base of one minute, the mean value of the four measurements given in Table 3 becomes 2006 CPM and the mean deviation becomes 8.6. This represents an error of only 0.4%—considerably smaller than that obtained with the smaller sample. Hence, we would certainly tend to believe that the results displayed in Table 3 are considerably more reliable than those in Table 2.

In the above experiment, we could have used the standard deviation instead of the mean deviation to measure the scatter of the measurements. Either method would give us a measure of the error. Of course, what represents an acceptable error in one situation might be totally unacceptable in another. The limits are set by, among other things, the features of the measurement process itself.

Chapter **5**

The Genetics of Populations and the Hardy-Weinberg Law[†]

In this chapter, we introduce some elementary aspects of population genetics. We shall attempt to explain the distributions of individuals with various genetic traits and we will attempt to predict how these traits will be distributed in subsequent generations. Before we begin this study, we first must make a rather extensive study of some of the elementary mechanisms of inheritance. Without this background, the full meaning of some of the most important results of population genetics cannot be appreciated. It is also essential to have some understanding of the physical basis for the genetic concepts that we shall employ, a matter toward which the following section is directed.

The Chromosomal Basis of Heredity

All biological organisms are single cells or aggregates of cells. Cells of higher organisms contain a region called the *nucleus* which in turn contains the *chromosomes*. When a dividing cell is stained and examined under a microscope, its chromosomes will, in general, be visible within the nucleus. It is now well established that the chromosomes are the carriers of the genetic determinants or *genes* and that their behavior is the basis for understanding the laws of inheritance.

In order to understand how the genetic characteristics of a cell are determined, it is necessary to examine the manner in which

---

[†] This chapter is based on material prepared by G. S. Kallin, H. O. Pollak, C. A. Grobe, Jr., et al. for a prototext on applied mathematics which was sponsored by the Teacher Training Panel of CUPM. The text will appear soon.

cells reproduce. The cells of every species of plant or animal contain a fixed number of chromosomes. Most cells in the organism reproduce by dividing into two identical cells (called *daughter* cells) in such a way that each of the daughter cells receives a set of chromosomes that is identical to that of the original cell. This process is called *mitosis*.

The *germ* cells (egg and sperm) are the exceptions to the above rule because they contain precisely half the number of chromosomes that the non-germ, or *somatic*, cells do. The process by which the germ cells are formed is called *meiosis*. Since an egg and a sperm each contain half the number of chromosomes, when they unite to form the fertilized ovum it will contain the correct number of chromosomes, having obtained half from the egg and half from the sperm. We shall examine both mitosis and meiosis in detail. Unless we specify to the contrary, we shall focus our discussion on an organism's somatic cells.

The cells of most organisms contain an even number of chromosomes because the chromosomes occur in pairs. Such cells are called *diploid*. One of each pair comes from the mother and the other from the father, as we shall see when we study the process of meiosis. For example, in the common fruit fly, *Drosophila melanogaster*, each cell contains 4 pairs of chromosomes. In humans, each cell contains 23 pairs. Corresponding pairs of chromosomes are called *homologues* or *homologous chromosomes*. Note that homologues are not necessarily identical.

Although a gene is known to consist of a certain area or region on a chromosome, we shall assume that each gene is just a position or locus on a chromosome. Usually, we shall associate with a given gene some physical characteristic of the organism such as its sex or eye color. In the case of a diploid organism, we specify that at a given locus on two corresponding chromosomes, alternative forms of a gene, called *alleles*, occur. We usually label them by letters. So, for example, an individual having the gene denoted by "A" on both corresponding chromosomes is labeled AA. We call such an individual *homozygous* or a *homozygote*. On the other hand, if an individual has the gene denoted by "A" on one chromosome and the gene denoted by "a" on the corresponding chromosome, then we denote the individual by Aa. Such an organism is called a *heterozygote*. (See Figure 5–1.) If, for example, an individual has

Homozygote Homologous chromosomes Heterozygote

Figure 5-1

genes A and a at the corresponding loci on two homologous chromosomes, we say that the individual's *genotype* is Aa.

We now proceed to examine the process of mitosis. We consider, as an example, an organism with two pairs of corresponding chromosomes in each cell. The first pair we label C_1 and c_1; the second pair we label C_2 and c_2. (See Figure 5-2.) Prior to reproduction via mitosis, each of the chromosomes duplicates itself.

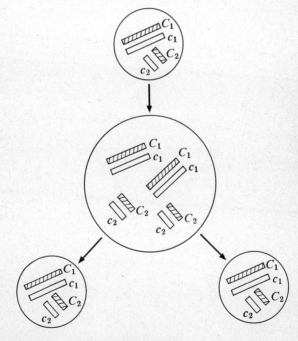

Figure 5-2

Hence, the cell contains two identical copies of C_1, two identical copies of c_1, and so on; i.e., the cell contains eight chromosomes. The cell then divides into two cells, each of which contains four chromosomes—one each of C_1, c_1, C_2, and c_2. Hence, each of the daughter cells is identical to the original cell.

The process of meiosis is slightly more involved. In the first step, each chromosome duplicates itself, so that the cell will contain two identical copies of each chromosome, as is the case in mitosis. But now, the cell divides into two cells in such a way that one cell contains both copies of C_1 while the other contains both copies of c_1; also one cell contains both copies of C_2 and the other contains both copies of c_2. Suppose, for the sake of argument, that one cell contains two copies of C_1 and two copies of C_2 while the other contains two copies of c_1 and two copies of c_2. (See Figure 5–3.) In the next step, each of these cells divides into two cells in the following way: Each of those coming from the cell that contains two copies of both C_1 and C_2 obtains one copy of C_1 and one copy of C_2; each of those coming from the cell that contains two copies of both c_1 and c_2 obtains one copy of c_1 and one copy of c_2. The four cells thus produced are germ, or reproductive, cells. Note that each contains exactly half as many chromosomes as does a somatic cell. Thus, when two germ cells combine to form the fertilized ovum, it will contain the correct number of chromosomes—in this case, four.

As a simple illustration of the way in which the genes combine to give an individual some particular trait, let us consider the way in which the sex of a human (and of most mammals) is determined. As stated above, each human cell contains 23 pairs of chromosomes, one pair of which determines, among other things, whether the individual is male or female. In examining the cells under a microscope, it was noted that if the cell comes from a male, the two corresponding chromosomes, or homologues, looked quite different. One of the pair is customarily denoted by "X" and the other by "Y." So, with respect to this pair of chromosomes, the male is denoted by XY, and is, therefore, a heterozygote. On the other hand, if the cell comes from a female, both of the corresponding chromosomes are the same. In fact, they are both X. So, with respect to this pair of chromosomes, the female is designated XX and is, therefore, a homozygote.

The Chromosomal Basis of Heredity

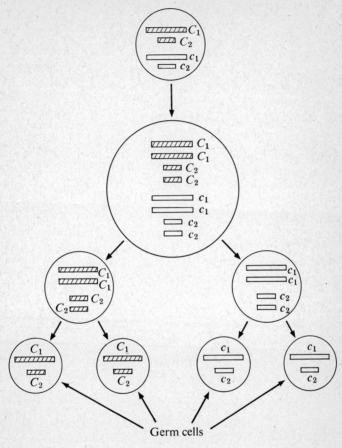

Figure 5-3

Now, let us consider how these chromosomes that determine an individual's sex are arranged in the individual's germ cells. As can be seen by examining Figure 5-4, each of the female's germ cells (the eggs) contains one X chromosome. (Remember that each of them also contains 22 other chromosomes. Since these do not influence the sex of the individual, we ignore them for the time being in order to keep the model as simple as possible.) Also, as can be seen from the figure, half of the male's germ cells contain an X chromosome and half contain a Y chromosome.

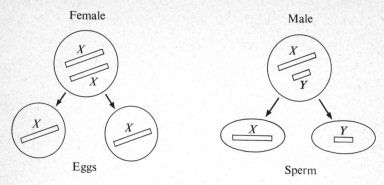

Figure 5-4

Now, suppose that one of the sperm unites with an egg to form the fertilized ovum. If, for example, a sperm containing and X chromosome fertilizes one of the eggs, then the ovum will contain two X chromosomes, having obtained one from the sperm and the other from the egg. Hence, the individual will be female. On the other hand, if a sperm containing a Y chromosome fertilizes one of the eggs, then the ovum will contain the X chromosome it obtained from the egg and the Y chromosome it obtained from the sperm. Therefore, the individual will be male.

Mendelian Inheritance

Modern genetics has its origins in a paper published in 1865 by the Austrian monk Gregor Mendel. Mendel worked with well-defined, discrete differences between pea plants. He asked, for example, what would happen if he crossed smooth peas with wrinkled peas. Would all of the offspring be somewhere between smooth and wrinkled? Or would half of them be smooth and half wrinkled? Mendel discovered that, in fact, all were smooth. The property of being wrinkled had disappeared. Mendel had crossbred to obtain this first generation. Thereafter, the peas reproduced by self-pollination; that is, the eggs of a given flower were fertilized by the sperm carried in the pollen of that same flower. Now, Mendel observed that a curious phenomenon had occurred. The trait which had been lost in the first generation appeared again in the second. That is, while all of the first generation were smooth, some of the second generation were wrinkled. Mendel counted the number

Mendelian Inheritance

of smooth peas and the number of wrinkled peas and discovered that the ratio was very nearly 3:1.

Mendel found that the same sort of behavior occurred if he crossed yellow peas and green peas. In the first generation of crossbreeds, all of the peas were yellow. If he then allowed the crossbreeds to breed by self-pollination, he found that some of the second generation peas were yellow and some were green. Furthermore, there were almost precisely three times as many yellow peas as green peas.

Mendel then asked: What is the result of crossbreeding peas which are smooth and yellow with those which are green and wrinkled? As before, the first generation were all yellow and smooth. But the second generation contained not only peas that were yellow and smooth, but also ones that were smooth and green, some that were yellow and wrinkled, and still others that were wrinkled and green. Furthermore, the ratio was approximately 9:3:3:1, respectively.

We propose to construct a model that will help to explain the above phenomena. Let us consider, first, the case of crossing yellow peas with green peas. We assume that each germ cell from a yellow pea carries a certain color gene, call it R, and that each germ cell from a green pea carries the allele which we shall denote by r. Therefore, when a germ cell containing an R unites with a germ cell containing an r, each cell will have genotype Rr. Because Mendel observed that all such individuals were yellow, we must conclude that the r allele is masked by the R, i.e., an individual with genotype Rr looks exactly like one with RR. In this case, we say that R is *dominant* and that r is *recessive*.

Next, Mendel allowed this first generation, that is, the peas with genotype Rr, to reproduce only by self-pollination. In order to explain why there are three times as many yellow peas as green peas in the second generation, let us observe that each germ cell from a first generation crossbreed will contain one of the alleles, i.e., it will contain either an R allele or an r allele. Now, what is the genotype of the offspring? There are four possibilities, as can be seen from Figure 5–5. If an egg containing an R allele is fertilized by a sperm containing an R allele, the offspring will have genotype RR. If the egg contains an R allele and the sperm contains an r allele, or conversely, the offspring will have genotype Rr. Finally, if both egg and sperm contain an r allele, then the offspring will have genotype rr.

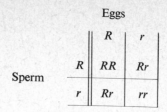

Figure 5-5

An individual with genotype RR will be yellow. So also will an individual with genotype Rr. But a pea with genotype rr will be green. Since germ cells containing the R gene combine randomly with those containing the r gene, our model suggests that, on the average, the results of such a breeding should produce 3 times as many yellow peas as green peas.

Let us make our model more sophisticated. Can we construct a model that will predict the 9:3:3:1 ratio between peas that are yellow and smooth, yellow and wrinkled, green and smooth, and green and wrinkled, respectively? In order to do so, let R denote the gene associated with yellow, S denote the gene associated with smoothness, r denote the gene associated with greenness, and s the gene associated with wrinkledness. So each germ cell from a pea that is pure yellow and smooth will have an R and an S gene; each germ cell from a pea that is pure green and wrinkled will contain an r and an s gene. Hence, the genotype of each of the crossbreeds is $RrSs$. Each germ cell of such a crossbreed will contain either an R or an r gene *and* an S or an s gene. What are the possibilities?

Again, we construct a table, Figure 5–6. This time, there are

	Eggs			
	RS	Rs	rS	rs
RS	$RRSS$	$RRSs$	$RrSS$	$RrSs$
Rs	$RRSs$	$RRss$	$RrSs$	$Rrss$
rS	$RrSS$	$RrSs$	$rrSS$	$rrSs$
rs	$RrSs$	$Rrss$	$rrSs$	$rrss$

Sperm (on left side)

Figure 5-6

sixteen (not all different) possibilities for the genotypes of the offspring. For example, if an egg cell containing the R and S genes unites with a sperm containing R and S genes, then the genotype of the offspring will be $RRSS$. If an egg containing an r and an S gene unites with a sperm containing an r and an s gene, then the genotype of the resulting offspring is $rrSs$. Since S is dominant over s, the offspring will be green and smooth.

In general, we note that any offspring which contains both an R and an S gene will be yellow and smooth. A pea whose cells contain an R gene but no S gene will be yellow and wrinkled; a pea whose cells contain an S gene but no R gene will be green and smooth. Finally, a pea whose cells contain neither an R gene nor an S gene will be green and wrinkled. Note that on the average, only one pea in 16 will be pure yellow and smooth, i.e., each of its cells will contain two R genes and two S genes. Since the germ cells combine in a random way, we see that the table constructed in Figure 5–6 does indeed lead us to conjecture that the ratio should be 9:3:3:1.

Our model may also be used to explain certain so-called "sex-linked" phenomena. In an historic experiment involving the common fruit fly, *Drosophila melanogaster*, the American biologist Thomas Hunt Morgan discovered a white-eyed fly among a batch of normal red-eyed flies. This fly, a male, whose white eyes were the result of a mutation, or spontaneous change, in the gene that controls eye color, was bred to a normal red-eyed female. All of the offspring had red eyes. Clearly, the gene responsible for red eyes is dominant over that responsible for white eyes. Then Morgan permitted these flies to breed with one another. As Mendel, and one of our above models, predicts, 75% of the second generation had red eyes and 25% had white eyes. But each of the white-eyed flies was male. Since the number of flies involved in the experiment was over 4,000, it couldn't be merely a coincidence that each of the white-eyed flies was male. Can we modify our model so that it can predict such behavior?

The determination of the sex in a fruit fly is analogous to that in a human. There is a pair of chromosomes which determines the sex. If each cell contains an identical pair, that is, if each cell contains two X chromosomes, the individual is female. If each cell contains one X chromosome and one Y chromosome, then the individual is male. Now, how do we account for the white-eyed males? We note, upon referring to the previous model, that each of the female's eggs contains one X chromosome while approximately half of the male's sperm contain an X chromosome and half contain a Y chromosome. Let's assume that the Y chromosome contains no gene which controls eye color, but that the X chromosome does. Let R denote the dominant gene for red eyes and let r denote the recessive gene for white eyes. So we assume that each of the white-eyed male's sperm which contains an X chromosome also carries an r gene. On the other hand, we assume that each of the red-eyed female's eggs, all of which necessarily contain an X chromosome, also carries the dominant gene, R, for red eyes. Now, what is the result of the mating? There are two possibilities. If a sperm carrying an X chromosome and, by assumption, an r gene, fertilizes one of the female's eggs, then the ovum will contain two X chromosomes, thereby guaranteeing that the individual will be female, and both an R gene and an r gene, thereby guaranteeing that the individual will have red eyes. The other possibility is that a sperm carrying a Y chromosome will unite with one of the female's eggs. In that case, the ovum will contain both an X and a Y chromosome, thereby guaranteeing that the individual is male, and the X chromosome will contain an R gene, thereby guaranteeing that the individual is red-eyed. Therefore, all offspring, regardless of sex, will have red eyes. See Figure 5–7.

Now what is the result of breeding two of these individuals together? First we examine the genetic makeup of all possible germ cells. Every egg necessarily contains an X chromosome. As can be seen from Figure 5–8, if the X chromosome came from the mother, then it contains an R gene. If it came from the father, then it contains an r gene. Next, let us examine the sperm cells of a male. Some will

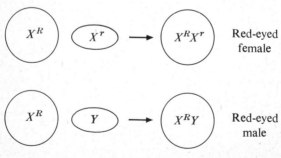

Figure 5-7

Mendelian Inheritance

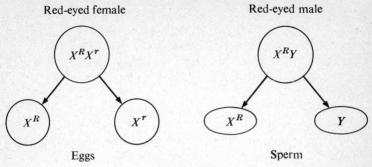

Figure 5-8

contain an X chromosome; others will contain a Y chromosome. Since the X chromosome came from the male's mother, it must contain an R gene. The Y chromosome, which necessarily came from the male's father, contains no gene that affects eye color.

Now, let's examine the genotypes which result from the random mating of two such individuals. There are four possibilities, as shown in Figure 5–9. As we can see from the figure, all individuals

		Eggs	
		X^R	X^r
Sperm	X^R	$X^R X^R$	$X^R X^r$
	Y	$X^R Y$	$X^r Y$

Figure 5-9

with white eyes will be males, although not all males will have white eyes, of course.

The above model is relevant to certain sex-linked disorders in human beings. If R represents the gene that permits a human to distinguish between red and green and if r represents its defective allele, then the model explains why mothers who are not themselves color-blind may pass the defect on to their sons. A similar explanation may be given for hemophilia.

The same point of view may also be utilized in order to explain the occurrence of another human disorder. In some parts of the world, particularly in the west of Africa, up to 25% of the population have abnormally shaped red blood cells called sickle cells. It is known that they have this abnormal shape because the hemoglobin of such people has a component which is different from that in people with normally shaped cells. The persistence of this abnormality in some parts of the world is a major problem for population genetics.

This abnormality can be explained if we assume that the relevant component of the hemoglobin in each individual is controlled by a pair of genes, one on each of a pair of homologous chromosomes. We shall denote these genes by R and r. If a person is of genotype RR, then he is normal. If he is of genotype Rr, then he will exhibit the sickle cell trait. If he is of genotype rr, then he will exhibit a more severe version of the sickle cell trait called *anemia*.

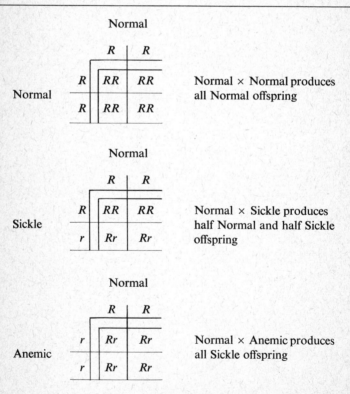

Figure 5-10

Since a parent contributes one of his genes at random to the offspring, the pair of genes received by the offspring is the combination of a random choice of one gene from each of his parents. The possibilities are shown in Figure 5–10.

We have, so far, considered only genes which occur in pairs. However, many genes occur in more than just two different forms. In addition to a dominant and a recessive gene, there may be one or more intermediate genes. The alleles are inherited in such a way that a particular individual receives any two of the genes and never more than two.

An important characteristic in man that is caused by *multiple alleles* is the blood type. In this example, there are three genes which control the formation of the blood corpuscles—*A*, *B*, and *O*. (We

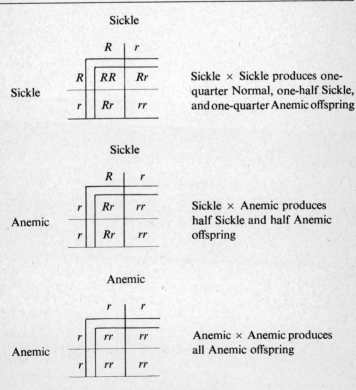

Figure 5-10 (cont.)

denote the genes by italic type in order to distinguish them from the related blood type.) *A* and *B* are each dominant to *O*, but neither *A* nor *B* is dominant with respect to the other. So, there are four blood types—A, B, AB, and O. In Figure 5-11 we tabulate the various blood types against the genotype.

Blood Type	Genotype	
A	*AA*	*AO*
B	*BB*	*BO*
AB	*AB*	
O	*OO*	

Figure 5-11

In Figure 5-12 we tabulate the possible blood types of a child as a function of the blood types of his parents. For example, the only way that two people, each with A blood, can produce a child with O blood is for both parents to be of genotype *AO* and for the recessive *O* genes to have occurred in both the egg and sperm that combined to form the child.

Parents	Children Possible	Children Not Possible
A and A	A or O	B or AB
A and B	A, B, AB, or O	
A and AB	A, B, or AB	O
A and O	A or O	B or AB
B and B	B or O	A or AB
B and AB	A, B, or AB	O
B and O	B or O	A or AB
AB and AB	A, B, or AB	O
AB and O	A or B	O or AB
O and O	O	A, B, or AB

Figure 5-12

Population Genetics

In the previous sections, we have examined some of the mechanisms by which hereditary traits are transferred from one generation to the next. We dealt almost exclusively with individuals. We now wish to turn our attention to the study of large groups of individuals, that is, populations. The purpose of this study is twofold. First, we will attempt to examine the genetic structure of natural populations, that is, the distributions of the genotypes within the population. Second, we will attempt to trace and predict the distribution of genotypes in future generations. The latter is, of course, the evolutionary process.

In order to accomplish these purposes, we will construct various models for populations and we will attempt to analyze them. While our models usually have to be simplified in order to be tractable, they must, in order to be valid, be based on an understanding of the mechanisms of inheritance.

A *population*, from the point of view of a geneticist, is a group of individuals who are able to mate with each other. We shall construct various models of a population, each slightly different and usually more complex than the last. In each model, we must be explicit concerning the following three basic ingredients:

1. *The definition of the genotypes to be studied. In particular we must, knowing the genotypes of the parent generation, define the way in which the parents' genes combine to form genotypes in the next generation.*
2. *The definition of the mating pattern with respect to these genotypes, i.e., the frequency with which each genotype occurs in the population.*
3. *The definition of each genotype's biological fitness. For example, we shall have to state the probability that a given genotype will live long enough to pass its genes on to the next generation.*

Two additional ingredients of a population model are mutation and migration. If they occur, they are prescribed in terms of the probability that a given genotype migrates or mutates. For our purpose, a *mutation* is defined as an instantaneous transition of one gene into another. Under natural circumstances, mutations occur rarely. We shall assume, unless specifically stated to the contrary, that neither mutation nor migration occurs. We *do* assume an infinite population.

In the first models, we consider a population where all individuals mate at the same time and are then completely replaced by their offspring. This is the so-called *nonoverlapping generations* model. It is not as crude a model as one might, at first glance, suspect, for there are many organisms whose populations behave in precisely this way. Most annual plants are an example.

In such a model, time changes are discrete and each unit corresponds to a given generation. In other words, such a model ignores the age structure of the population.

Random Mating. The Hardy-Weinberg Law

We shall consider a model which satisfies the following assumptions:

1. There are precisely two alleles, R and r, at a particular locus.
2. The three possible genotypes, RR, Rr, and rr, mate randomly.
3. All individuals reproduce at the same rate and all have equal probabilities of surviving to reproduce. That is, we assume that all three genotypes have equal biological fitness.

Now, suppose in a given generation, call it G_0, that the three possible genotypes, RR, Rr, and rr, have population frequencies u_0, $2v_0$, and w_0, respectively, where $u_0 + 2v_0 + w_0 = 1$. This means that the ratio of the genotypes RR to Rr to rr is $u_0 : 2v_0 : w_0$. By assumption, the three genotypes mate at random. We now ask: In the next generation, denoted by G_1, what is the proportion of the three possible genotypes to each other? Since, by assumption, the three genotypes belonging to G_0 mate randomly, we can see, by referring to Figure 5–13, that the proportions are $(u_0 + v_0)^2 : 2(u_0 + v_0)(v_0 + w_0) : (v_0 + w_0)^2$. So, if we let

$$u_1 = (u_0 + v_0)^2,$$
$$v_1 = (u_0 + v_0)(v_0 + w_0),$$
$$w_1 = (v_0 + w_0)^2,$$

then the ratio of genotypes RR, Rr, and rr to one another in G_1 is just $u_1 : 2v_1 : w_1$.

G_0			G_1		
			Offspring		
Male	Female	Frequency	RR	Rr	rr
RR	RR	u_0 u_0	u_0^2		
RR	Rr	u_0 $2v_0$	$u_0 v_0$	$u_0 v_0$	
RR	rr	u_0 w_0		$u_0 w_0$	
Rr	RR	$2v_0$ u_0	$u_0 v_0$	$u_0 v_0$	
Rr	Rr	$2v_0$ $2v_0$	v_0^2	$2v_0^2$	v_0^2
Rr	rr	$2v_0$ w_0		$v_0 w_0$	$v_0 w_0$
rr	RR	w_0 u_0		$u_0 w_0$	
rr	Rr	w_0 $2v_0$		$v_0 w_0$	$v_0 w_0$
rr	rr	w_0 w_0			w_0^2

Frequency of RR individuals in G_1:

$$u_0^2 + 2u_0 v_0 + v_0^2 = (u_0 + v_0)^2$$

Frequency of Rr individuals in G_1:

$$2(u_0 v_0 + u_0 w_0 + v_0 w_0 + v_0^2) = 2(u_0 + v_0)(v_0 + w_0)$$

Frequency of rr individuals in G_1:

$$v_0^2 + 2v_0 w_0 + w_0^2 = (v_0 + w_0)^2$$

Figure 5-13

We can, of course, continue this process. If we define u_2, v_2, and w_2 by the equations

$$u_2 = (u_1 + v_1)^2,$$

$$v_2 = (u_1 + v_1)(v_1 + w_1),$$

and

$$w_2 = (v_1 + w_1)^2,$$

then in the next generation, G_2, the ratio of genotypes RR, Rr, and rr to one another is $u_2 : 2v_2 : w_2$. And so on.

Let us consider an example. Suppose that in G_0, we have

$$u_0 = \frac{1}{10},$$

$$2v_0 = \frac{7}{10},$$

$$w_0 = \frac{2}{10}.$$

Hence

$$u_0 : 2v_0 = \frac{1}{7}$$

and

$$2v_0 : w_0 = \frac{7}{2}.$$

Then using the formulas for u_1, v_1, and w_1 given above, we find that

$$u_1 = \frac{81}{400},$$

$$v_1 = \frac{99}{400},$$

$$w_1 = \frac{121}{400}.$$

Referring to the formulas for G_2 given above, we calculate that

$$u_2 = \frac{81}{400},$$

$$v_2 = \frac{99}{400},$$

and

$$w_2 = \frac{121}{400}.$$

Hence, $u_1 = u_2$, $v_1 = v_2$, and $w_1 = w_2$. And, if we calculate the frequency of the three genotypes in G_3, we find that $u_1 = u_2 = u_3$, $v_1 = v_2 = v_3$, and $w_1 = w_2 = w_3$.

Random Mating. The Hardy-Weinberg Law

Is this phenomenon peculiar to this particular example, or is it true in general?

Exercise: Let $u_0 = 1/4$, $2v_0 = 1/2$, and $w_0 = 1/4$. Compute u_i, v_i, w_i, $u_i : 2v_i$, and $2v_i : w_i$ for $i = 1, 2$, and 3.

Exercise: The same as the above with $u_0 = 1/100$, $v_0 = 0$, and $w_0 = 99/100$.

The exercises should convince you that the frequency of the three genotypes remains fixed after the initial generation, that is, $u_{i+1} = u_i$, $v_{i+1} = v_i$, and $w_{i+1} = w_i$, for $i = 1, 2, \ldots$. This fact was discovered independently in 1908 by the British mathematician G. H. Hardy and a German physician, W. Weinberg.

Exercise: Hardy's paper may be found in *Science*, **28** pp. 49–50. Read it!

Theorem (Hardy-Weinberg): *Let u_0, $2v_0$, and w_0 be any three nonnegative numbers such that $u_0 + 2v_0 + w_0 = 1$ and such that no two are zero. Let*

(1)
$$u_{i+1} = (u_i + v_i)^2,$$
$$v_{i+1} = (u_i + v_i)(v_i + w_i),$$
$$w_{i+1} = (v_i + w_i)^2,$$

for $i = 0, 1, 2, \ldots$. Then, for $i = 1, 2, 3, \ldots$,

$$u_{i+1} = u_i,$$
$$v_{i+1} = v_i,$$

and

$$w_{i+1} = w_i.$$

Proof: Suppose that neither u_0, v_0, nor w_0 is zero. We will show that the ratios $u_i : 2v_i$ and $2v_i : w_i$ are independent of i for $i = 1, 2, 3, \ldots$. This, coupled with the fact that $u_i + 2v_i + w_i = 1$, will prove the theorem.

First, we observe from equations (1) that

(2)
$$v_i^2 = u_i w_i$$

for $i = 1, 2, \ldots$.

In particular, equation (2) gives

$$v_1^2 = u_1 w_1.$$

If we add $u_1 v_1$ to both sides of the above equation, we have

$$u_1 v_1 + v_1^2 = u_1 v_1 + u_1 w_1,$$

or

$$v_1(u_1 + v_1) = u_1(v_1 + w_1).$$

This implies that

$$\frac{u_1}{2v_1} = \frac{u_1 + v_1}{2(v_1 + w_1)}$$

$$= \frac{(u_1 + v_1)^2}{2(u_1 + v_1)(v_1 + w_1)}$$

$$= \frac{u_2}{2v_2}.$$

That is to say, $u_1 : 2v_1 = u_2 : 2v_2$. In a similar way, we can show that

$$v_1^2 = u_1 w_1$$

implies $2v_1 : w_1 = 2v_2 : w_2$. But since

$$u_1 + 2v_1 + w_1 = u_2 + 2v_2 + w_2 = 1,$$

it follows that $u_1 = u_2$, $v_1 = v_2$, and $w_1 = w_2$.

In order to complete the proof, we simply proceed inductively. That is, from equation (2), we have

$$v_2^2 = u_2 w_2.$$

This implies that $u_2 : 2v_2 = u_3 : 2v_3$ and $2v_2 : w_2 = 2v_3 : w_3$. From this, it follows that $u_2 = u_3$, $v_2 = v_3$, and $w_2 = w_3$. And so on. This proves the theorem.

Exercise: Suppose that exactly one of the numbers, u_0, v_0, or w_0 is zero. Prove the Hardy-Weinberg theorem in this case.

Now, let us examine the model from a slightly different point of view. We are interested in determining the frequency of the two

Random Mating. The Hardy-Weinberg Law

genes R and r in each of the populations $G_0, G_1, G_2, \ldots$. Let p_i denote the frequency of R in G_i and let q_i denote the frequency of r in G_i. Obviously, $p_i + q_i = u_i + 2v_i + w_i = 1$.

Recall that in the parental generation G_0, the genotypes RR, Rr, and rr occur with frequencies u_0, $2v_0$, and w_0, respectively. Therefore,
$$p_0 = u_0 + v_0$$
and
$$q_0 = v_0 + w_0.$$

In G_1, the genotypes RR, Rr, and rr occur with frequencies

$$(u_0 + v_0)^2 \quad 2(u_0 + v_0)(v_0 + w_0) \quad \text{and} \quad (v_0 + w_0)^2,$$

respectively. Therefore,

$$\begin{aligned} p_1 &= (u_0 + v_0)^2 + (u_0 + v_0)(v_0 + w_0) \\ &= p_0^2 + p_0 q_0 \\ &= p_0(p_0 + q_0) \\ &= p_0. \end{aligned}$$

By a similar argument, we can easily show that
$$p_{i+1} = p_i$$
and
$$q_{i+1} = q_i.$$

Therefore, under the assumptions of our model, the gene frequency in each generation is the same as that of the preceding generation. Hence, for $i = 1, 2, \ldots$, the distribution of the genotypes RR, Rr, and rr in G_1 is p_0^2, $2p_0 q_0$, and q_0^2, respectively. That is, for $i = 1, 2, \ldots$,
$$u_i = p_0^2$$
$$v_i = p_0 q_0$$
$$w_i = q_0^2.$$

Hence, in the case of this simple model, the genotype frequencies in any generation G_1 are specified once the gene frequencies are known in the parental population G_0. This is, of course, just another way of stating the Hardy-Weinberg Theorem.

Random Union of Gametes

In this section, we analyze a model that is a slight variant on the one contained in the previous section. Specifically, we assume:

1. There are precisely two alleles, R and r, at a particular locus.
2. The three possible genotypes RR, Rr, and rr shed their gametes in large numbers into a "pool" and new individuals are formed by choosing pairs of gametes at random out of this pool.
3. All individuals have equal probabilities of surviving to reproduce. That is, we assume that all three genotypes are of equal biological fitness.

Again, we assume that the parental generation G_0 contains the genotypes RR, Rr, and rr with frequencies u_0, $2v_0$, and w_0, respectively, where $u_0 + 2v_0 + w_0 = 1$. As before, the gene frequencies for R and r are given by

$$p_0 = u_0 + v_0$$

and

$$q_0 = v_0 + w_0.$$

Hence,

R and R are chosen with probability p_0^2 to give RR.

R and r
  or   are chosen with probability $2p_0q_0$ to give Rr.
r and R

r and r are chosen with probability q_0^2 to give rr.

This process may be represented as in Figure 5-14. Therefore, the

		Male gametes	
	Freq.	R	r
		p_0	q_0
Female gametes R	p_0	p_0^2	$p_0 q_0$
r	q_0	$p_0 q_0$	q_0^2

Figure 5-14

distribution of genotypes in this model is exactly the same as that in the previous model.

The above model can help us to analyze the frequency with which certain recessive genes occur in a population. In human beings, for example, the gene r for albinism is recessive to the gene R for normal skin pigmentation. Surveys have shown that the frequency of albinos, who have genotype rr, is about 1 in 20,000. That is (we drop the subscripts), $q^2 = 1/20{,}000$. Hence, $q = 1/141$. So, $p = 1 - q = 140/141$. Now, what is the frequency of people who are heterozygous for albinism? The answer:

$$2pq = 2 \times \frac{140}{141} \times \frac{1}{141}.$$

This comes out to be approximately 1/70; i.e., although only one person in 20,000 is an albino, one person in 70 carries the recessive gene for albinism.

In addition to the human blood types described earlier in this chapter, there are three other types which are usually denoted by M, MN, and N. These blood types are determined by two alleles denoted by M and N. (Again, we italicize the gene in order to distinguish it from the related blood type.) The blood type as a function of the genotype is shown in the following table:

Blood type	M	MN	N
Genotype	MM	MN	NN

The frequencies of these blood types in the U.S. population have been observed. These frequencies are given in the following table:

Blood type	M	MN	N
Observed frequency in U.S. population	0.292	0.496	0.212

Now, how well do the observed frequencies agree with those predicted by the Hardy-Weinberg Theorem?

If we let u, $2v$, and w be the frequencies of the three possible

genotypes, MM, MN, and NN, respectively, then from the above table,
$$u = 0.292,$$
$$v = 0.248,$$
$$w = 0.212.$$
So
$$p = u + v = 0.540,$$
$$q = v + w = 0.460.$$
So
$$p^2 = 0.2916,$$
$$2pq = 0.4968,$$
$$q^2 = 0.2116.$$

Hence, the frequencies predicted by the Hardy-Weinberg Theorem are in very good agreement with those obtained by observation.

PM '72

D. C. Heath and Company

QUAN.	CATALOG NO.	D E S C R I P T I O N
1	075275	INTRO MATHEMATICAL BIOLOGY

NO CHARGE

SENT TO YOU WITH THE COMPLIMENTS OF D.C. HEATH AND COMPANY.
IF YOU DESIRE FURTHER INFORMATION,
PLEASE CALL US TOLL FREE AT:
    1-800-225-1388
    (IN NYC 800-225-1388
    IN MA. 1-800-842-1211)

OR CONTACT YOUR DISTRICT MANAGER:
LOREN LE JEUNE
HOLIDAY PLAZA OFFICE BLD
1050 NORTHGATE DRIVE-SUITE 320
SAN RAFAEL, CALIF. 94903

FORM 16-0152 REV. 11/71

Chapter 6

The Applications of the Computer to Biology

Within the last twenty-five years, digital computers have evolved with enormous rapidity. This has brought about profound changes in all of the sciences, including biology. Biologists are now able to manipulate and analyze enormous amounts of very complex data. In this chapter, we shall describe, in a general way, the working of a digital computer and how a user communicates with the machine. Then, we shall indicate some of the ways in which such a computer can be of use to a biologist. It is not our intention to instruct the reader in computational techniques, for these are outside the scope of this small book. Instead, we refer the reader to the list of references.

The Modern Digital Computer

In reality, the largest computers are merely very complex machines that are capable of performing only the simplest of elementary arithmetic operations. They have, as we have remarked, the ability to manipulate astronomical quantities of data in an extremely short time. The machine can be instructed to perform a sequence of operations where, at each step in the sequence, the calculation that it performs may depend upon the answers obtained in each of the previous steps of the sequence. This feature permits the machine to perform incredibly complicated calculations.

Schematically, we may think of our computer as consisting of five components, namely, the *input*, the *output*, the *memory*, the *arithmetic unit*, and the *control unit*. We illustrate this in Figure 6–1. The solid arrows indicate the directions in which information may be transferred. The dotted arrows represent either the way in which instructions may be passed from one unit to another or the way in

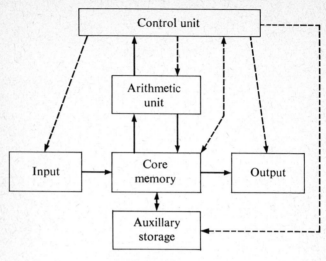

Figure 6-1

which control of the various components is carried out. The input is a device, often a card or tape reader or a teletypewriter, that allows us to feed our data and instructions into the machine. After the machine has performed the calculations that we have asked it to perform, it spits out the answer on output. This device may be a high-speed line printer or it may be the same teletypewriter which we used to input the program and data into the machine. Or, it may be a graphical plotter or a cathode-ray device, both of which display the answers to us. Some output devices are connected back to the experimental apparatus that generated the data in the first place. The best example, perhaps, of this sort of arrangement is the use of computers in X-ray diffraction studies of large molecules. The results of one measurement are fed into the computer, analyzed, and then, depending upon the result of the analysis, are fed back to the experiment where the computer then directs the taking of a second set of measurements. The process can be repeated as many times as is necessary. In this way, the computer directs the experiment and analyzes the results in the virtual absence of human direction.

The memory is an electronic device for storing data and answers in the machine. It may take the form of signals that are stored on magnetic tape or it may consist of information that is stored on punch cards or on paper tape. In a large, modern machine, the memory is, in fact, several memories. There is a relatively small memory, called the *core memory*, that is located in close proximity to the arithmetic unit and the control unit. While the capacity of this core memory is usually quite small in comparison to the total memory of the computer, information may be retrieved from it in a matter of somewhere around a millionth of a second. The so-called access time to the rest of the memory is considerably longer, but the amount of information that may be economically stored there is considerably greater. Such devices may take the form of a magnetic tape or disc drive or punch cards or paper tape.

The arithmetic unit is the electronic device which actually performs the calculations which are required. The brain of the computer is the control unit. This is the device that controls the sequence in which the operations are actually carried out by the machine. For example, it may be programmed to retrieve a certain piece of data from a specified location in the memory, perform some operation on it, and then store the results somewhere in the memory for future reference. Hence, the control unit is programmed to control the flow of information to and from the memory, to specify what arithmetic operations are to be performed and in what order, and to control the input and output of information from the computer.

The Program

The *program* is a set of instructions that the computer is directed to follow in order to carry out a certain calculation. The preparation of a program, then, requires first that we decide exactly what the computer must do in order to perform the calculation that we want. Then, we must translate this program into a language that the computer can understand. Let us consider these two problems in that order, i.e., first we will illustrate how we formulate a program that will accomplish some task which we have in mind.

As an example, we consider the famous Fibonacci sequence given by 1, 1, 2, 3, 5, 8, 13, 21, ... How, precisely, is this sequence generated? If we denote the nth element of this sequence by a_n—so $a_1 = 1$, $a_2 = 1$, $a_3 = 2$, $a_4 = 3$, and so on—we see that $a_3 = a_1 + a_2$, $a_4 = a_2 + a_3$, $a_5 = a_3 + a_4$, and so on. That is, starting with the third element of the sequence, each element is the sum of the previous two. In symbols, then, we can define the Fibonacci sequence by the equations,

$$a_1 = 1,$$
$$a_2 = 1,$$
$$a_{n+2} = a_n + a_{n+1}, \quad \text{if } n = 1, 2, 3, \ldots.$$

Now, suppose that we wish to write a program that will allow us to generate on a computer all of the elements of the Fibonacci sequence which are less than or equal to 1000. Before we set up a flow chart for this process, we first must establish some conventions.

If we draw a rectangular box as shown below, then we are

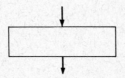

instructing the computer to execute the instructions contained in the box and then to pass on to the next box in the chart. If we draw an oval box as shown below, then we place an assertion inside the box.

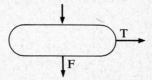

If the assertion is true, then we instruct the machine to perform some task; if the assertion is false, then we instruct it to do something else.

In Figure 6–2 we have a flow chart that will permit a computer to calculate and print out the elements of the Fibonacci sequence that are less than or equal to 1000.

Note that the flow chart contains what is called a "loop," that is, an instruction which effectively requires that the machine perform the same sort of operation over and over again. It would, of

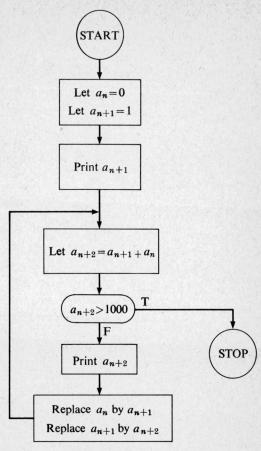

Figure 6-2

course, go on performing this iteration *ad infinitum* were it not for the decision box, which instructs the computer to exit from the loop as soon as a_{n+2} exceeds 1000. Note, also, that it is necessary to tell the computer to start and to stop.

After a reasonable method has been devised to carry out a program, then the problem becomes to communicate this strategy to the machine itself. This is the second, or "semantic," phase of programming. More precisely, the program, together with the relevant data, must be inserted into the memory of the computer in such a way that it will direct the machine to carry out the proper steps in the correct sequence. Back in the dark ages of digital computers, that is, fifteen years ago, a user had to communicate with his computer using what is called "machine language." This requires an extremely detailed knowledge of the internal workings of the machine. It involves directing each step of the calculation and deciding where each and every piece of information is to be stored in the memory. For people who know how to use it, machine language is still the most flexible and efficient means of programming, but it is the tool of the specialist.

The proliferation of the use of the computer into all fields has necessitated the invention of languages that may be more easily mastered by the novice, although, as in so many areas of human endeavor, there is a question of whether the chicken or the egg came first. At any rate, languages have been devised that resemble, for example, English, and that allow the relative amateur to converse easily with the computer. One such example is the computer language called BASIC. A computer that is capable of accepting BASIC uses a teletypewriter for the input device. Such a teletypewriter usually has only capital letters. Therefore, it is essential that we distinguish between the letter "O" and the number "zero." We do this by denoting the latter by $\emptyset$; hence, the number ten is written as 1$\emptyset$.

In order to illustrate how easy it is to write a program using BASIC, we give, below, a program which will compute the average of the four numbers 1, 2, 5, and 8. The program is:

```
1∅    LET A = (1 + 2 + 5 + 8)/4
2∅    PRINT A
3∅    END
```

Note that each statement in the program is preceded by a number. This tells the machine in which order the instructions are to be carried out. We then type RUN and the machine will type out 4, the answer. We can make the program fancier, as follows:

```
2     DATA      1, 2, 5, 8
5     READ      B, C, D, E
```

        10 LET A = (B + C + D + E)/4
        20 PRINT A
        30 END

When the machine encounters the READ statement, it will automatically assign to each of the variables in the READ statement the numbers in the DATA statement in the order in which they appear in the DATA statement. So, in our example, the computer will assign 1 to the variable B, 2 to the variable C, and so on.

The advantage of this program over its predecessor is that it may be more easily used to compute the average of *any* four numbers. In order to instruct the machine to compute the average of 10, 3, 21, and -4, for example, we simply replace line 2 above by

        2 DATA   10, 3, 21, -4.

All other lines in the program will remain the same and the computer will then proceed to crank out the average of these four numbers.

In order to translate complicated formulas such as

$$V(S, I) = \frac{V}{1 + \frac{K}{S}\left(1 + \frac{I}{1.2}\right)}$$

into BASIC (this equation occurs in the study of enzyme rates), we write

        80 DEF FNV(S, I) = V/(1 + K/S*(1 + I/1.2)).

The symbol "*" means to multiply. Clearly, some convention must be established regarding the order in which the operations are to be carried out. Rather than dwell upon this further, we refer the reader to any of several manuals on BASIC.

As a final example, we list a program which will compute the binomial coefficient. Recall from our chapter on probability that we defined the binomial coefficient by the equation

$$B = \binom{R}{S} = \frac{R}{1} \cdot \frac{R-1}{2} \cdot \frac{R-2}{3} \cdot \ldots \cdot \frac{R-S+1}{S}$$

where R and S are positive integers with $R \geq S$. The program for computing $\binom{R}{S}$, where $R = 5$ and $S = 3$ is:

```
 5  DATA 5, 3
10  READ R, S
15  LET B = 1
20  IF S > 0 THEN 35
25  PRINT B = 1
30  GO TO 70
35  LET N = R + 1
40  LET D = 0
45  LET N = N − 1
50  LET D = D + 1
55  LET B = B*N/D
60  IF D < S THEN 45
65  PRINT B
70  END
```

Several comments are in order regarding the above program. Line 20 is appropriately enough called an "if-then" statement. If the condition $S > 0$ is satisfied, then the computer jumps to statement 35; if the condition is not satisfied, then the computer proceeds to the next statement (in this example, 25). The statement contained in 50 may be confusing if you are encountering it for the first time. It is to be interpreted as follows: You have a variable that you call D. Statement 40 says to assign D the value 0. Then statement 50 says to replace the old value of D, namely 0, by a new value of D, namely 1. The next time the machine encounters 50, it will replace the old value of D, now equal to 1, by a new value, namely 2. And so on. The same idea works in statements 45 and 55.

In addition to being easy to use, there is another feature of BASIC that endears it to both the novice and the experienced programmer. For example, in the program that we gave above for computing the average of 1, 2, 5, and 8, if we had replaced statement 10 by

10 Let A = (1 + 2 + 5 + 8/40)

the machine, when instructed to compute the number, will reply something like

ILLEGAL FORMAT IN 1∅.

In other words, the machine has the ability to spot simple errors in our program. But until a machine is devised which can think, it won't detect errors in our logic.

Languages such as BASIC require that a program be stored in the memory of the computer. Such a program translates the statements that we input on the teletypewriter into machine language. This means, in effect, that a "translator" must be kept in the computer's memory. For this reason, such languages make inefficient use of the central processing unit.

In addition to BASIC, there are a number of other languages, such as FORTRAN and ALGOL, the latter most widely used in Europe. You can become quite proficient in the use of any one of these in just a few hours of concentrated study. We now turn our attention to a general consideration of two of the ways in which computers may be used by the biologist.

The Use of Computers in Ecology

There are two principal ways in which computers can be utilized to study ecology. First, computers may be used for data sorting, correlation, and statistics. Because ecology deals with such exceedingly complex systems of organisms, it yields data in bewildering richness. Under the best of circumstances, the analysis of this data can be so time consuming as to be practically impossible if one has to analyze it using nothing more sophisticated than, say, a desk calculator. Modern digital computers have changed this. Now, more and more ecological data are being obtained in a form that can be fed directly into a computer for analysis. The availability of machines that are capable of handling enormous amounts of data has removed much of the barrier to considering highly complex systems but, needless to say (or, perhaps it isn't), it has not eliminated the necessity for the researcher to know precisely what he is attempting to learn from his ecological data. On the contrary, the computer has placed more of a burden than before on the ecologist to formulate his questions in a very precise way. One suspects, perhaps, that this secondary effect is at least as important to the study of ecology as the more obvious one of allowing the ecologist to analyze heretofore unassailable quantities of data.

The analysis of models is the second and, perhaps, more interesting way in which a large digital computer can be of assistance to the ecologist. Much work has recently been done on the problem of reducing ecological problems to problems in mathematics, that is to say, to the construction of mathematical models. Because problems in ecology are themselves so complex, the models based on the ecological system are also exceedingly complicated. Hence, the analysis of these models has depended in great part on modern digital computers.

For example, one might be interested in the fluctuation of a certain population of organisms as a function of the food supply and the predation. At first glance, this might seem quite simple. As the food supply increases, it is certainly reasonable to expect that the population will increase. On the other hand, an increase in the number of predators will tend to decrease the population. Hence, one might obtain a set of mathematical expressions that serve to describe how the population under study varies with the food supply and the number of predators. But even such a simple-sounding model might, in fact, turn out to be quite complicated if one requires, as one should, that the model be capable of predicting results that can actually be observed in the field. For example, an increase in the population of a certain species often heralds an increase in the populations of the species that prey upon the given organisms. It has been documented, for instance, that the population of snowy owls in the arctic increases during the times when lemmings are in abundant supply. And, another consideration: Not only is the population dependent upon the food supply, but the food supply per individual is usually dependent upon the population. For example, allowing too many sheep to graze on a certain tract of land may kill the grass. And finally, even if the food were in unbounded supply, the territory available to the population is usually not. For instance, it has been observed in some rats that overcrowding can cause a decrease in the population. Hence, constructing a model, which at first glance seemed childishly simple, turns out to be in reality very complex.

In the actual construction of any model that is to represent a complicated system, it is safe to say that a considerable period of trial and error ensues before the model begins to predict what can actually be observed. In other words, it is usually necessary to make a number of adjustments in the model before it becomes a reasonable one. Therefore, it is very helpful if one can quickly obtain the effect that varying one of the parameters in a model will have on the model. For this reason, a computer that can, for example, plot the predicted population of a certain species is very useful. If you want to change the assumptions regarding, say, the change in the population of the predator as a function of the change in the population of the organism under study, then you can quickly see the change that the model predicts in the population under study.

Computers in Physiology and Biochemistry

As in the case of ecology, computers are used in the physiological sciences to analyze experimental data and to evaluate models. For example, one can extract from data the kinetic constants that relate the rate of reaction to the reactant concentration. Here, the tactic is to fit the constants to the experimental data in such a way as to minimize differences between the experimental curve and the appropriate enzyme rate equation.

In addition, computers have proved useful in the analysis of results that deal with very complicated metabolic and physiological problems. One such example is the flow of material through branching pathways and across cellular membranes. In many cases, these processes are fairly well understood only on the cellular level. Computers appear to be of great service in extending our understanding to the indescribably more complex case of a multicellular organism.

Finally, computers are of great importance in the construction and evaluation of physiological and biochemical models. The reader will understand that the relatively simple approach to enzyme rates taken in the third chapter represents something of an ideal case and that an accurate description of the much more involved cases that exist in, for example, intact organisms, can be extremely difficult. As is true in ecology, the construction of alternate models is frequently of great assistance in discovering what is really going on.

For one thing, a complex biochemical event is not easily resolved into simple component parts without making a number of what may turn out to be unwarranted assumptions. The validity of these assumptions can often be quickly tested by analyzing the resulting model on a computer. At the same time, the examination of alternative models may suggest experiments that might not otherwise occur to the investigator.

While it is true that any model that can be analyzed on a computer can also, in principle, be analyzed via manual means, the time required for the latter may make it a practical impossibility. And even if it is possible, the time required may be so great as to severely limit the number of alternative models that it is practical to evaluate.

One can quickly appreciate the utility of a computer-borne model by considering a very simple case of a hypothetical metabolic pathway consisting of several reactions as shown diagrammatically by

$$A \overset{1}{\rightleftharpoons} B \overset{2}{\rightleftharpoons} C \overset{3}{\rightleftharpoons} D \ldots$$

Imagine, furthermore, that each of these individual reactions is well understood; that is, the values for maximum velocity and Michaelis constant, K_m (see Chapter 3), are known. We simply want to know if the pathway is as drawn; that is, is it a nonbranching series in the order shown? This may be accomplished by comparing the rate of flow through the sequence (which is measurable in the laboratory) with the predicted rate under a variety of conditions. We obtain the predicted rate by solving a set of enzyme rate equations that take into consideration the fact that the product of one reaction is the substrate for the next. We might begin by making the steady-state assumption; i.e., we assume that the concentrations of the intermediates remain constant with time. This would require that the rate of formation and breakdown of each intermediate be the same and would therefore greatly simplify things. Under these conditions, we could, perhaps, make our calculation with or without the use of a computer. Hence, the great utility of the computer would become apparent only when we discover that our calculated rate was far removed from the measured one. We would then want to abandon the steady-state assumption and introduce, instead, possible branches in the chain or alter the properties of some of the enzymes. One would probably want to (or have to) alter more than one variable at a time. This would make the computer look better and better.

As a final bit of compelling evidence that computers are of service in evaluating enzyme models, let us contemplate the case of noncompetitive inhibition (which we will not describe in any detail). Imagine again that we wish to examine a model by varying such details as the various kinetic constants until values come out that conform to experimental data. Whether a computer is required for this is, perhaps, a matter of taste, but it is worth noting that the equation based on the most reasonable model (which, incidentally, includes several simplifying assumptions) takes the following form:

$$v = kp[ES]$$

$$= \frac{\dfrac{kpF_0[S]}{K_M}\left[1 + r\dfrac{k_1[I]}{k_1[I] + (ks[S] + k - s + k - 1)(1 + r)}\right]}{1 + \dfrac{[S]}{K_M} + \dfrac{[I]}{K_1} + \dfrac{[S][I]}{K_M K_1}}$$

$$\times \frac{1}{\left[1 + r\dfrac{k - 1 + k_1[I]}{k_1[I] + (ks[S] + k - s + k - 1)(I + r)}\right]}$$

Chapter 7

Some Applications of Advanced Mathematics

In the previous chapters, we have seen several examples of some of the ways in which elementary mathematics may help to provide insight into complicated biological problems. We hope that you have become convinced that a knowledge of elementary mathematics may considerably enhance your understanding of some sophisticated and complicated biology. We hope, on the other hand, that you have *not* become convinced that a smattering of elementary mathematics will suffice for the understanding of all complicated biological systems. Such is not the case. We urge you to learn as much mathematics as you can. Not only is it a beautiful subject; it can be useful as well.

We subscribe to the dogma which says, in effect, that you can never effectively use all of the mathematics that you know. Now, we don't wish to embark upon a philosophical discussion of the meaning of the word "know"; instead, we are simply attempting to suggest that in order to use calculus—just to choose an example—really effectively, you should know some advanced calculus. In short, the password is "overkill." Also, unless you possess foresight of a kind denied most mortals, you will never know exactly (or, probably, even remotely) what mathematics will prove useful to you in your investigations of biological phenomena. So learn all that you can.

In the remainder of this chapter, we will briefly treat some of the more advanced mathematics that is turning out to be of importance to biologists. We will end this book (or harangue) by assumnig that you have been won over. Hence, we will suggest a number of sources that will lead you to bigger and better things.

Calculus and Differential Equations

One of the reasons why calculus is so powerful a tool for the physical scientist is that it provides him with a neat way of describing how one variable will change with respect to some other variable. In a previous chapter, for example, we were concerned with the way in which a population of bacteria changed with respect to time. Recall that we hypothesized that one of the equations that we could use to describe this phenomenon is

$$\frac{dN}{dt} = kN,$$

$$N(0) = N_0.$$

We observed that if you know just a bit of calculus, you can write down the solution to these equations by inspection, namely,

$$N(t) = N_0 e^{kt}.$$

Unfortunately, this example is an unusually simple one. More often one encounters an equation that is not trivial to solve. In fact, computers are frequently used in order to obtain approximate solutions.

We can illustrate the way in which another complication arises as follows: Suppose that we are interested in studying the rate of change of concentration of some substance across a membrane. In particular, suppose that we are interested in knowing the concentration not only as a function of time but also as a function of the distance from the membrane. Hence, we study the concentration as a function of the two variables, time and distance. This may lead us to what is called a *partial differential equation*—a subject which can be extremely difficult.

Biologists are often interested in finding the maxima and minima of a function. For example, organisms or systems of organisms are frequently stable at certain minimum values of some variable such as energy. Hence, the search for such "singularities" is of importance. There are procedures, using calculus, that permit us to find such points.

The study of vectors, and the calculus associated with them, is becoming of increasing importance to the biologist. A vector represents a magnitude that is oriented in some given direction. For example, you can think of a force as being a vector. It is frequently represented as a line segment whose length is proportional to the magnitude of the vector and whose direction is given with respect to a certain fixed coordinate system. Vectors occur in many places in biology. The charge associated with the beating of a heart is one such example. There is a part of calculus which analyzes the way in which vectors change with time and position. A biologist who understands this subject will be in a position to better understand such conceptually difficult problems as, for example, the interactions between different substances that are flowing across a living membrane.

Algebra

Although you undoubtedly studied algebra at a very tender age, it continues to be of considerable interest to the modern research mathematician. It is primarily concerned with the mathematics of discrete systems.

Group theory, for example, has seen considerable application in the study of crystals. This has led, in turn, to important discoveries in protein and nucleic acid biochemistry. Perhaps the most famous outcome of this study is the construction by Watson and Crick of a model for double-stranded DNA.

Matrix algebra, also, is of considerable interest to the biologist. A matrix, for our purposes, is just a rectangular array of numbers such as

$$\begin{bmatrix} a_{11} & a_{12} & \cdots & a_{1n} \\ a_{21} & a_{22} & \cdots & a_{2n} \\ \vdots & & & \\ a_{m1} & a_{m2} & \cdots & a_{mn} \end{bmatrix}.$$

Each element in the array carries with it two subscripts. The first subscript denotes the row to which the element belongs and the second the column. Hence, a_{23} occurs in the second row and the third column. Matrices allow you to solve large collections of linear equations. They have been frequently applied to such diverse biological subjects as statistics, the kinetics of simultaneous cellular reactions, and the properties of neural networks.

Topology

Certain areas of mathematics, such as statistics, are applied to biology in extremely practical and immediate ways and can frequently be used by investigators who know little about their conceptual foundations. Other areas, by way of contrast, appear to be of use in biology only in the very general sense that they provide useful guides for thinking about biological questions. For example, the field of topology has to date produced very little of practical benefit to biological science, but it may, in fact, suggest some very fruitful approaches to present problems. Perhaps it also holds some promise of more concrete benefits in the future.

Topology, for our purposes, may be thought of as a subject that emphasizes what might be called the spatial relationships between objects. It is filled with such notions as surfaces and "closeness" and so, in a way, helps to give us geometrical insight. For example, certain biological membranes appear to contain enzymes on their surface as a part of their structure. Thus, the inner membrane of a cell organelle, the mitochondrion, is, in part, composed of a number of the enzymes associated with cellular energy metabolism. Of these enzymes, some appear to react, not with a substrate of low molecular weight, but with each other. One enzyme may do nothing more than pass an electron on to a second enzyme, with which it must be in contact. Now, given this sort of membrane, it is possible to ask some questions about the relations between enzymes. For example, it is clear that for the system of enzymes to function in an optimum fashion, in the simplest case, each enzyme should be in contact with a different enzyme with which it can react rather than be in contact with an enzyme that is identical to it. Given this restriction, we might well ask how many different enzymes could be included in such a system. If we do ask this, we are recasting in biological guise an old topological problem, namely, the four-color map problem. This asks whether or not every map on a plane can be colored with four colors without adjacent regions being the same color. (It is legitimate for two regions of the same color to meet at a point, but the two regions may not share an arc or line as part of their common boundary.) This question is not trivial, having been around (and unsolved) since 1852.

Although we cannot provide a solution to the four-color map problem (nor to the enzymic corollary), a considerable amount is known about such maps. This knowledge may provide insight into the properties of biological surfaces such as the one described. Indeed, a topological kind of thinking can be of considerable help in dealing with cell structure, in general, because cells are often composed of very complicated systems of continuous membranes. One may ask, for instance, if the mitochondrion is topologically spherical; that is, can it be stretched into a sphere without cutting any surfaces or altering any fundamental topological properties (as opposed to geometrical properties that are obviously altered by stretching)? In fact, the mitochondrion appears to be topologically two concentric spheres and a number of its most interesting properties result from this configuration.

Suggested Additional Readings

In the event that you may wish to pursue further the matter of mathematical approaches to biology, we list below several books which may be of interest. The list is necessarily incomplete, but the references given in these books will greatly extend the selection. We have purposely refrained from including works in biology and mathematics *per se*. Thus, if you desire to find out more about growth or enzyme structure, you would be well-advised to examine a general biology textbook and, perhaps, go on from there.

First of all, there are several books in the general area of biomathematics, all of which can be highly stimulating. For example, two classical works, recently reprinted in paperback form, are:

    LOTKA, A. J., *Elements of Mathematical Biology*. New York: Dover Publications, Inc., 1956

and

    RASHEVSKY, N., *Mathematical Biophysics: Physico-Mathematical Foundations of Biology*, Third Edition. New York: Dover Publications, Inc., 1960.

Newer books in the same area include:

    BAILEY, N. T. J., *The Mathematical Approach to Biology and Medicine*. New York: John Wiley and Sons, Inc., 1967

    SMITH, J. M., *Mathematical Ideas in Biology*. Cambridge: Cambridge University Press, 1968 (paperback)

and

    WATERMAN, T. H. and H. J. MOROWITZ, ed., *Theoretical and Mathematical Biology*. Waltham, Mass.: Ginn and Company, College Division, 1965.

Modern algebra appears to be of increasing significance and value to the biologist. There are two recent books dealing with this subject:

> NAHIKIAN, H. M., *A Modern Algebra for Biologists*. Chicago: University of Chicago Press, 1964

and

> SEARLE, S. R., *Matrix Algebra for the Biological Sciences*. New York: John Wiley and Sons, Inc., 1966.

Additional information about computers and programming is very easy to come by due to the growing list of texts and practical guides. The conceptual foundation of the computer is largely the discipline of information (or communication) theory whose birth could be said to coincide approximately with the publication of the following two books:

> WEINER, N., *Cybernetics*, Second Edition. Cambridge, Mass.: M.I.T. Press, 1961. (First Edition, 1948)

and

> SHANNON, C. E. and W. WEAVER, *The Mathematical Theory of Communication*. Urbana: University of Illinois Press, 1948.

The second makes, on the whole, more mathematical demands than the first. As the reader of Weiner's book can immediately tell, the study of information theory is of considerably more interest to biologists than that of computers alone. Two excellent books giving additional and more recent evidence of this connection are:

> ARHIB, M. A., *Brains, Machines and Mathematics*. New York: McGraw-Hill, Inc., 1964

and

> GEORGE, F. H., *Cybernetics and Biology*. San Francisco: W. H. Freeman and Co., 1964.

The latter of these is a paperback and is somewhat more elementary

Suggested Additional Readings

than the former. On a more practical level, there are many, many books on digital computation, one being

> GREEN, B. F., *Digital Computers in Research*. New York: McGraw-Hill, Inc., 1963

and there are numerous books on the various programming languages including the following, which introduces the language BASIC mentioned in Chapter 6:

> KEMENY, J. G. and T. E. KURTZ, *BASIC Programming*. New York: John Wiley and Sons, Inc., 1967.

The following is also of interest to the slightly more advanced computer user:

> ORGANICK, E. I., *A FORTRAN Primer*. Reading, Mass.: Addison-Wesley Publishing Company, Inc., 1963.

An interesting introduction to analogue computation is given in

> HARTLEY, M. G., *An Introduction to Electronic Analogue Computers*. London: Methuen, 1962.

A student wishing to know more about the mathematics of physiological and enzymatic processes would do well to examine

> BERNHARD, S. A., *The Structure and Function of Enzymes*. New York: Benjamin Company, Inc., 1968 (paperback)

or

> GOODWIN, B. C., *Temporal Organization in Cells*. New York: Academic Press Inc., 1963.

Likewise, a mathematical view of populations may be found in

> MACARTHUR, R. and J. CONNELL, *The Biology of Populations*. New York: John Wiley and Sons, Inc., 1966

and

> SLOBODKIN, L. B., *Growth and Regulation of Animal Populations*. New York: Holt, Rinehart and Winston, Inc., 1961.

Finally, there are a number of serial publications in the area of mathematical biology (of quite a range of sophistication). In addition, mathematical approaches are to be found in nearly all serious journals that deal with biological matters. To obtain an idea of what is being done at the present time in this field the reader would do well to examine current issues of any of the following:

Biophysical Journal
Bulletin of Mathematical Biophysics
Computers and Biomedical Research
Journal of Theoretical Biology
Mathematical Biosciences
Progress in Biophysics and Biophysical Chemistry.

Index

A

Active site, enzyme, 21
Advanced mathematics, 83
Alleles, 46, 51, 55, 57, 60, 66

B

BASIC, 74
Binomial coefficient, 39, 75
Binomial distribution, 38
Blood type, 57, 58, 67
Branch, 31
Branch values, 32

C

Catalyst, 23, 24
Chemostat, 14
Chromosomes, 45–55
Computer, 69
Computer arithmetic unit, 69, 71
Computer control unit, 69, 71
Computer core memory, 71
Computer input, 69, 70
Computer memory, 69, 70
Computer output, 69, 70
Computer program, 71
Counts per minute, 42

D

Death rate constant P, 17
Diploid cells, 46
Doubling time t_d, 12

E

Enzymes, 21–27
Enzyme inhibition, 30
Enzyme rates, 23–27, 75
Enzyme specificity, 21
Enzyme-substrate complex, 26
Equilibrium constant K_{eg}, 23
Escherichia coli, 5

F

Fibonacci sequence, 71–72
Frequency, F, of an event, 36–37

G

Genes, 45, 51, 54, 56–59, 65
Genotype, 47, 51, 53, 58–68
Germ cell, 46, 48, 51
Group theory, 85
Growth rate, 9, 16
Growth rate constant K, 9, 14, 17

H

Half-life, 41
Hardy-Weinberg Law, 60, 63
Hemophilia, 55
Heterozygote, 46, 48
Homologues, 46
Homozygote, 46, 48

K

K_m, Michaelis constant, 27–29, 80

Index

M

Machine language, 74
Matrix algebra, 85
Mean deviation, 40–43
Mean value, 35, 40–43
Median value, 35
Meiosis, 46, 48
Michaelis constant K_m, 27–29, 80
Mitosis, 46–48
Modifier molecule, 30
Mutation, 59

N

Network theory, 30
Nodes, 31
Nonoverlapping generations, 60

P

Partial differential equation, 84
Path, 31
Path values, 32
Population, 59
Population genetics, 45, 58
Probability, 33ff, 36–37

R

Rate constant k, 22–26, 32
Rates of chemical reactions, 21

S

Sickle cell anemia, 56
Somatic cells, 46, 48
Standard deviation, 41
Steady-state assumption, 24

T

Topology, 86

V

Variance, 41
Vector, 84–85